The Plasma Membrane: Models for Structure and Function

The Plasma Membrane:
Models for Structure and Function

B. D. Gomperts
Department of Experimental Pathology
University College Hospital Medical School
London, England

1977

ACADEMIC PRESS
LONDON   *   NEW YORK   *   SAN FRANCISCO

A Subsidiary of Harcourt Brace Jovanovich, Publishers

ACADEMIC PRESS INC. (LONDON) LTD
24–28 Oval Road
London NW1 7DX

U.S. Edition published by
ACADEMIC PRESS INC.
111 Fifth Avenue
New York, New York 10003

Library of Congress Catalog Card Number: 76 016968

ISBN: 0 12 289450 2

Printed in Great Britain by
T. & A. Constable Ltd, Edinburgh

Preface

MEMBRANE MODELS AND MODEL MEMBRANES

The study of membranes has recently become one of the most compelling topics for any student of biology, and a knowledge of at least the jargon that modern membrane science has thrown up is almost *de riguer* for anyone with pretensions of spanning the two cultures. No reader of *New Scientist* or *Scientific American* needs to be told that the words "fluid mosaic" refer to a generalised model of membrane architecture[1]; and for many, the mere mention of its ultimate authors, Singer and Nicolson (like Watson and Crick of yester-year, maybe?) conjures up a clear picture which takes them on a visionary ride to the forefront of scientific endeavour. The descriptive shorthand used to identify modern membrane science has been cleverly epitomised in a way which will surely be meaningful[2] (even when reproduced *sans mots*) to most readers of this book.

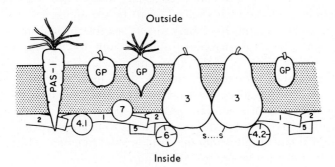

The trouble with such an easily memorable and pictorial approach for the student (or indeed, for anyone interested in cells) is that any such all-embracing description has to fail just as soon as it is applied to individual structures. Such is the nature of biological diversity. This is not the fault of those who first applied the simplifying epithet to complex forms, because the detailed descriptions out of which the final shorthand arose were probably valid. The fault lies in the frailty of our own eyes, ears and comprehension. We are, most of us, far too willing to substitute an

over-simplified concept for the knowledge and understanding generated by years of endeavour. (The havoc wrought by the too ready
acceptance of the "high energy phosphate bond" concept in the field of
bioenergetics is a warning that we in membrane science would do well to
heed[3].)

For these reasons I have tried to avoid the use of such simplifying
conceptual descriptions wherever possible, though in one or two places
they have crept in because to have avoided them entirely would have been
a form of censorship, and I have no wish to rewrite the history of our
science. The term "lipid bilayer" is also a piece of descriptive shorthand
which hides a wealth of wisdom: I use it unashamedly throughout, but I
hope that in this case I have provided sufficient detail so that the reader
has some real understanding of what lies behind the words.

Some of the reasons for the current excitement in the field of membrane
science are to be found in the great technical developments in instrumentation during the fifties and sixties, which allowed more and more specific
questions to be asked of more and more complex structures. Thus the
techniques of X-ray diffraction, calorimetry, nuclear magnetic resonance
and the various forms of optical spectroscopy have all been applied to the
paraffins, to the soaps, to phospholipids, to the "simpler" biological
membranes, and now to many tissues *in situ*. A contrasting approach to the
description of any structure is obtained by fixation and electron microscopy,
for here, instead of signals averaged over space and time, it is possible to
observe isolated and unusual details down to the limit of resolution. The
last twenty years have seen enormous strides in both instrumentation and
in the handling of the biological material. In a number of ways, the advances in understanding have been made possible by advances in tech-
nology, and it is clear that one could attempt to describe membrane
structure simply in terms of the information yielded by the various forms
of instrumental measurement: indeed, this perfectly valid analytical
approach has already been used on a number of occasions by others in
introducing biological membrane structure.

As an alternative, I have attempted to write what is essentially a synthetic
description of biological membrane structure and function. The techniques
of modern instrumentation are still important, though I have decided to
impose a somewhat arbitrary limit, and even a cursory look at the text will
show that all references to the "extrinsic probes" (notably spin-labels and
fluorescence) are absent. This is certainly not to belittle their importance,
for, at times, certain information gained from the use of the extrinsic
probes has led to new insights, particularly in the dynamic aspects of
membrane structure. It is my belief, however, that these techniques are
too readily applied by those who are insufficiently prepared to use them.
Elementary fluorescence measurements are far too easy to make, and
invariably yield a welter of data which in the long run can prove too hard

to convert into useful information. These techniques are in truth best left to those who possess the fundamental understanding required to interpret their meaning; and that really goes for the reader too. Of course, this can be said about any technique, but we have to start somewhere. Every student of science should have some idea of what is meant by the state of matter, and by phase change, and so the information gained from calorimetric measurements is immediately accessible. When we get to X-rays and NMR, we are on more dangerous ground, but to take the case of NMR, let us in the first place simply accept it as a spectrum. If we now use it to glean information just as one might use the absorption spectra of the reduced and oxidised cytochromes (i.e. we neglect its theoretical basis) there should be no obstacles and we can use it as a springboard to greater understanding.

It is truly astonishing to find in a textbook of 1957 the sentence: "There is the same general lack of positive evidence of the functional roles of phospholipids . . . as was noted for sterols. . . ." Certainly, it would be hard now, following on the developments in the chemistry of membrane components, initially of the lipids and latterly of the specialised proteins peculiar to membranes, to overstress the importance of the phospholipids in membrane structure and function. However, if we take the quotation to its conclusion ". . . with the additional difficulty of accounting for the great range of lipids: lecithins, three components of the classical 'cephalins' plasmalogens, sphingomyelins, cerebrosides and gangliosides"[4] we find, that in some ways we may not have travelled too far. One way in which the importance of individual membrane components has been most clearly illustrated has been through the development of model membranes: synthetic structures of defined composition having just some of the properties of biological membranes; and it is with this background that the subject of biological membrane structure and function is introduced.

In the preparation of this book I have relied on the help, advice and assistance of many friends. Some of these "friends" are, in truth, not known to me personally, but to any reader of the book it should not prove hard to determine who they are. I am indebted to Mr Durward Lawson who generously provided a number of previously unpublished electron microscope photographs. The text has been read at various stages by Drs Roger Dean, Patrick Riley, Clare Fewtrell, Robert Simmons and Martin Raff. Dr Paul Mueller read a section of Chapter 3 and Dr John Foreman read a section of Chapter 4. All these people gave excellent guidance, not all of which did I accept. All errors of fact, omission and of judgement are therefore mine.

B. D. GOMPERTS

October 1976

REFERENCES

1. Singer, S. J. and Nicolson, G. L. (1972). The fluid mosaic model of the structure of cell membranes. *Science*, **175**, 720.
2. Steck, T. L. (1974). The organization of proteins in the human red blood cell membrane. *J. Cell Biol.* **62**, 1.
3. Banks, B. E. C., (1969). Thermodynamics and biology. *Chemistry in Britain*, **5**, 514.
4. Lovern, J. A. (1957). "The Chemistry of Lipids of Biochemical Significance." Methuen, London.

Contents

1 | Models for Membrane Structure

WHY START WITH MODELS?

Biological structures are composed of the molecules of chemistry, and the rules that control their reactivity are the same as those which control reactivity in chemistry. But there is one special feature of biology which is absent from chemistry and the physical sciences, and that special feature is *function*. All biological structures have functions, and all the biological structures that we recognise have been developed (in a Darwinian sense) to optimise their functional capacities. This generalisation can be extended all the way from the gross exterior appearance of the beast, down to the molecular intermediates of the metabolic process. So it is with membranes. Membranes are composed of molecules; these must obey the rules of physics and chemistry. But membranes are also developed to perform functions.

One way of approaching the question of membrane function is to consider how biological membranes differ from (or resemble) "membranes" which lack any functional capacity.[1, 2, 3] These are the model membranes; inert structures, composed generally, but not invariably, of molecules derived from biological membranes into which it is now possible to insert "functional" centres (often toxic substances of microbiological origin) and so reproduce some of the most highly developed forms of biological transport phenomena in a selected and simplified manner. As examples of this, the generation of action potentials, redox gating, control of the rate of osmotic water movement by hormones, and light triggered conductance changes have all been described in model membranes suitably treated with substances of biological origin. In this approach we can begin to see what it is that is special about biology, for without the description of the model membrane we would have no baseline: we would have no way of knowing

1

exactly what are the problems which biological development has had to overcome.

Some of the model membranes described in this chapter (the liposomes) probably resemble the primitive prebiotic membraneous structures which first allowed the distinction to be made between inside and outside.[4] For this reason it can be said that the approach of discussing model membranes before biological structures is chronologically appropriate.

MATERIA MEMBRANARUM

As a preliminary to constructing model membranes we should consider the nature of the materials from which they should be made. The presence of fatty materials in the membranes of biological cells was predicted by Overton in 1899,[5] long before the techniques for isolating them and analysing their chemistry had been developed. He timed the onset of plasmolysis of plant root cells (i.e. the contraction of the membrane-limited cell within the fixed space bounded by the rigid plant cell wall) after placing them in hyperosmotic sucrose solutions containing various penetrating solutes. Extensive and maintained plasmolysis observed with a microscope indicates slow penetration of the solute; rapid inhibition indicates rapid penetration. In general the penetrating solutes were shown to be those which are readily soluble in ether and fatty oils, and it was inferred that the protoplasm is bounded by a fatty barrier. Overton even went so far as to venture the suggestion that the proposed fatty barrier might contain lecithin and cholesterol. Other major components of biological membranes are proteins and carbohydrates.

PHOSPHOLIPIDS

Membrane lipids can be separated from the non-lipid material by extraction with a mixture of chloroform, methanol and water. The lipids can then be separated into their various classes by chromatographic procedures. Of biological membranes, that of the mammalian red blood cell is the most easily prepared free from contamination by other cellular structures: it is, indeed, often regarded (unwisely) as being synonymous with the very notion of "membrane" (see Chapter 2, p. 97). Its lipid composition is probably typical of the plasma membranes of other mammalian cells. It contains about 10% of glycolipid, and the remainder consists of phospholipid and cholesterol in about equal proportions, and it is with these materials that most model structures are built.

We consider now the chemistry of the phospholipids especially with regard to a number of special features which might affect their role as the central structural components of membranes. Phospholipids are termed *amphipathic* (αμφιπαθεια, both loving) substances. Examination of the general structure of the glycerol phospholipids illustrates why (Fig. 1).

general structure of the glycerol phospholipids

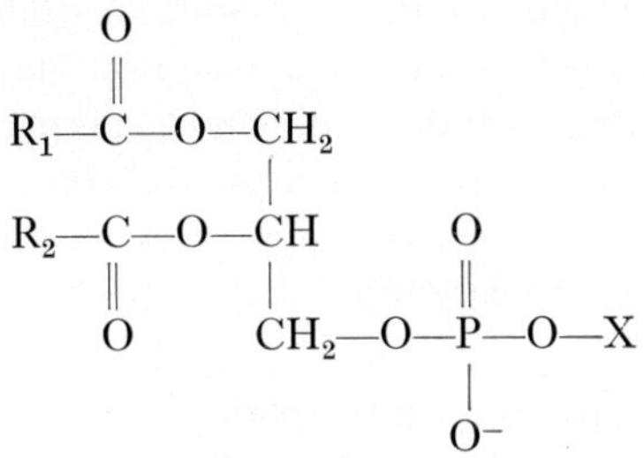

$$
\begin{array}{l}
\quad\quad\quad\quad O \\
\quad\quad\quad\quad \parallel \\
R_1-C-O-CH_2 \\
\quad\quad\quad\quad\quad | \\
R_2-C-O-CH \quad\quad O \\
\quad\parallel\quad\quad\quad | \quad\quad\quad \parallel \\
\quad O \quad\quad CH_2-O-P-O-X \\
\quad\quad\quad\quad\quad\quad\quad\quad | \\
\quad\quad\quad\quad\quad\quad\quad\quad O^-
\end{array}
$$

		net charge at neutral pH
X = —H	phosphatidic acid	negative
$= -CH_2CH_2-\overset{\overset{\displaystyle CH_3}{\displaystyle |}}{\underset{\underset{\displaystyle CH_3}{\displaystyle |}}{N^+}}-CH_3$	phosphatidyl choline (PC) (lecithin)	neutral
$= -CH_2CH_2-N^+H_3$	phosphatidyl ethanolamine (PE)	neutral
$= -CH_2CH_2-\overset{\overset{\displaystyle N^+H_3}{\displaystyle |}}{\underset{}{COO^-}}$	phosphatidyl serine (PS)	negative
$= -CH_2\overset{\overset{\displaystyle OH}{\displaystyle |}}{CH}-CH_2OH$	phosphatidyl glycerol (PG)	negative
cardiolipin is diphosphatidyl glycerol, and has phosphatidic acid esterified at the 1- and the 3-hydroxyl groups of glycerol		negative

FIG. 1. Structures of the glycerol phospholipids.
A simplified nomenclature is used in this text to describe the chemistry of the phospholipids. This is best illustrated by example: diC 18:0 phosphatidyl choline (or diC 18:0 PC) describes a phosphatidyl choline in which both hydrocarbon chains have 18 carbon atoms and no double bonds, i.e. distearoyl phosphatidyl choline. 18:0/18:1 PE describes a phosphatidyl ethanolamine in which R_1 has 18 carbon atoms and is fully saturated (stearoyl) and R_2 also has 18 carbon atoms but has one double bond (oleyl), i.e. 1-stearoyl,2-oleyl-phosphatidyl ethanolamine: the position of the double bond in the hydrocarbon chain is not specified. The descriptive title "lecithin" is retained to describe the mixtures of phosphatidyl cholines found in purified extracts from natural sources, i.e. the title "egg yolk lecithin" refers to the phospholipids purified as phosphatidyl cholines from egg yolk.

AMPHIPATHY

R_1 and R_2 are long-chain fatty acids ranging in length from 12 to 24 carbon atoms. Whilst R_1 is generally saturated, R_2 (depending on its length) may contain up to six olefinic double bonds. A single phospholipid isolated from a homogeneous source may contain a great variety of fatty acids, varying in both chain length and unsaturation. This is well illustrated by the analysis (by gas-liquid chromatography after hydrolysis) of a representative sample of egg yolk phosphatidyl choline which had the following composition[6]:

Chain	% (w/w of total fatty acid)	Chain	% (w/w of total fatty acid)
16:0	26·2	20:2	2·8
16:1	2·0	20:4	5·4
18:0	15·1	20:5	2·8
18:1	31·9	22:5	2·8
18:2	12·2	22:6	4·4
18:4	2·8		

Notice that the fully saturated components account for just over 40% of the total fatty acid in this case. It is likely that these are exclusively located in the R_1 position. The phosphoester "head group" confers a hydrophilic character at the end of a molecule which otherwise has the properties of a simple fatty substance. Thus the hydrocarbon chains search for a lipophilic environment while the head group searches for an aqueous environment. It is readily seen that phospholipids are endowed with interfacial characteristics, with the tails looking one way, the heads the other. Figure 1 illustrates the structures of some of the more familiar phospholipids. Phosphatidic acid is not truly a membrane lipid, but is obtained by hydrolysis of phosphatidyl choline with phospholipase D (cabbage juice). The phosphoric acid group esterified to glycerol confers a negative charge on this substance. Phosphatidyl choline (PC) is a neutral zwitterion, and is the predominant neutral phospholipid of most mammalian plasma membranes. Phosphatidyl ethanolamine (PE) is a neutral zwitterion at physiological pH: above pH 9, the amino group will dissociate to make the molecule negatively charged. Phosphatidyl serine (PS) carries a net negative charge at neutral pH, due to the presence of a positive amino group and two negatively charged groups: carboxyl and phosphate. Phosphatidyl glycerol (found in the membranes of plants and lower organisms) and cardiolipin (found in mitochondrial inner membranes) carry the simple negative charges of the phosphate groups.

SPHINGOLIPIDS

The generic structures of some sphingolipids are shown in Fig. 2. The common feature here is the base sphingosine:

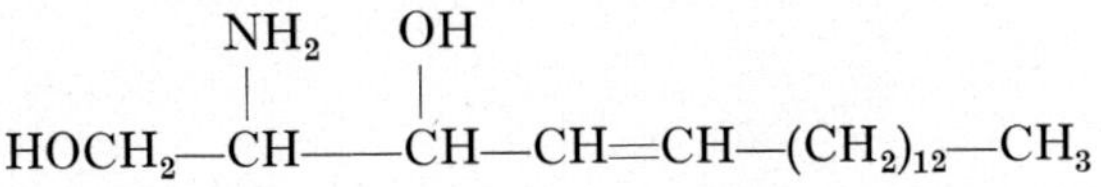

$$\overset{\displaystyle NH_2}{|}\quad\overset{\displaystyle OH}{|}$$

$$HOCH_2\text{—}CH\text{——}CH\text{—}CH\text{=}CH\text{—}(CH_2)_{12}\text{—}CH_3$$

This is best thought of as a straight chain alcohol (C_{18}) having a unique double bond and substituted with amino and hydroxyl groups at carbon atoms 2 and 3. The sphingolipids carry only a single fatty acid residue, and this is linked via an amide linkage at C_2 of the sphingosine.

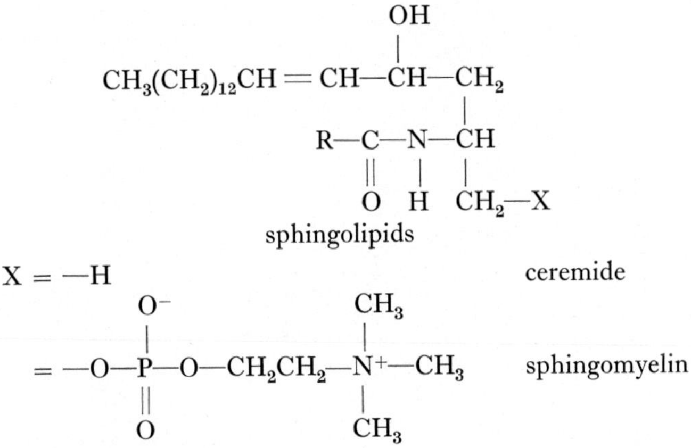

$$\overset{\displaystyle OH}{|}$$

$$CH_3(CH_2)_{12}CH = CH\text{—}CH\text{—}CH_2$$
$$R\text{—}C\text{—}N\text{—}CH$$
$$\overset{||}{O}\ \ \overset{|}{H}\ \ \overset{|}{CH_2}\text{—}X$$

sphingolipids

$X = $ —H   ceremide

$$= \text{—}O\overset{\displaystyle O^-}{\underset{\displaystyle O}{\overset{|}{\underset{||}{P}}}}\text{—}O\text{—}CH_2CH_2\text{—}\overset{\displaystyle CH_3}{\underset{\displaystyle CH_3}{\overset{|}{\underset{|}{N^+}}}}\text{—}CH_3 \qquad \text{sphingomyelin}$$

$=$ simple sugar of polysaccharide   cerebroside
which may be N-acetylated

$=$ polysaccharide containing   ganglioside
N-acetylneuraminic acid (NANA,
i.e. acetylated sialic acid)

FIG. 2. Structures of the sphingolipids.

The fatty acid of typical sphingolipids is fully saturated, C_{18} (stearic) and C_{24} (lignoceric) predominating. Sphingomyelin, the chief lipid of myelin (and also an important component of the human red cell membrane), is derived from ceremide by the esterification at C_1 of phosphoryl choline which gives the head group of sphingomyelin the same zwitterionic properties as phosphatidyl choline. The cerebrosides and gangliosides are derived by glycosidic linkage at C_1. The more complex carbohydrate structures attached to these molecules are antigenic in quality, and it is possible that they play a role in cell recognition, but of their precise role, little is known. (The chemistry and structures of the phospholipids, sphingolipids and related materials is excellently presented in reference 7.)

CHOLESTEROL

The structures of cholesterol, and some related sterols are illustrated in Fig. 3. Cholesterol is a major component of the plasma membrane of all

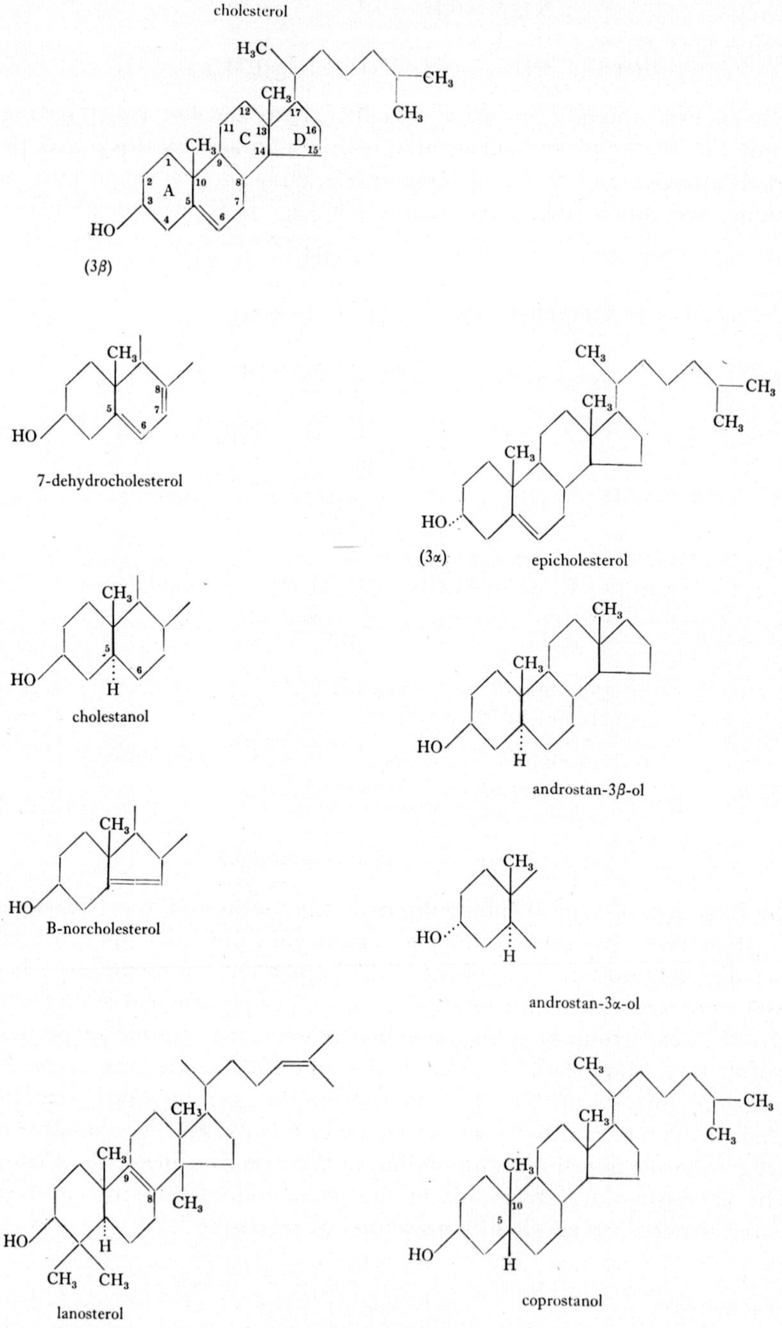

cholesterol

(3β)

7-dehydrocholesterol

epicholesterol

cholestanol

androstan-3β-ol

B-norcholesterol

androstan-3α-ol

lanosterol

coprostanol

FIG. 3 (a). The structure of cholesterol and some related sterols.

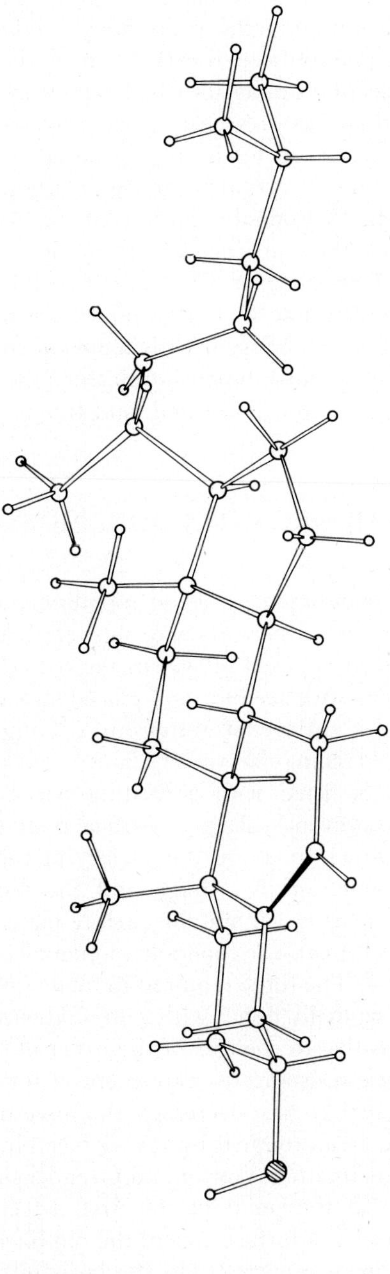

FIG. 3 (b). A drawing of a molecular model of cholesterol.

mammalian tissues, but it is substantially absent from certain intracellular organelles, such as the mitochondrial inner membrane. In most plasma membranes it is present in an equimolar ratio with the phospholipids. Like the phospholipids, it has amphipathic properties, deriving from the presence of the hydroxyl group in β conformation (i.e. lying in the equatorial position at C_3) and the remainder of the molecule which is hydrophobic. In association with phospholipids, these features ensure that cholesterol is aligned so as to present the hydroxyl group to the region occupied by the polar head groups, and the hydrophobic section to the region occupied by the esterified fatty acid residues. The hydrophobic portion of cholesterol is divided into two distinctive sections: the four rings which comprise the steroid nucleus, and the aliphatic tail attached at C_{17}. The fused rings of the steroid nucleus form a rigid plate-like structure, on which are mounted methyl groups at C_{10} and C_{13} (as can be seen in the model of cholesterol illustrated in Fig. 3), which confer a third dimension to this otherwise flat molecule. The aliphatic tail at C_{17} is uncomplicated, and free to flex and rotate.

MOLECULAR ORGANISATION: MONOLAYERS

The most important experiment in all of membrane science was that of Gorter and Grendel,[8] who, fifty years ago, showed that the phospholipids might be organised as an opposed bilayer in the red cell membrane. They extracted red blood cells with acetone, and placed measured samples of the lipid extract on to the surface of water, in a Langmuir trough. The Langmuir trough is a rectangular vessel about 1 cm deep filled with an aqueous medium. It is fitted with a floating barrier having platinum brushes at the ends to prevent leakage of surface material past the barrier. The applied extract which floats on the surface of the water is confined between three fixed sides of the trough and the floating barrier. The barrier is now moved so as to confine the surface material within an ever-decreasing area, and ultimately a coherant monomolecular film forms on the surface of the water. The force required to maintain this surface film is measured. This was originally done[9] with a torsion wire and optical lever system which was sensitive to changes of the order of 0·01 dyne cm^{-1}.

Initially there is little resistance to movement of the barrier, but as the area occupied by the surface film decreases, the force needed to move the barrier increases. The area occupied by the surface film is plotted against the force required to maintain it. Gorter and Grendel showed that the area occupied by the film at the point of the first detectable increment of pressure was about twice the surface area of the red blood cells from which the lipid extract had been prepared. On the basis of this evidence, they suggested that the lipids in red cell membranes are arranged as a continuous double molecular layer, a hypothesis which has been at the centre of every discussion of membrane structure from that day to this.

The monolayer technique has been used to find out about the packing and surface electrostatic properties of amphipathic substances. It underwent a kind of renaissance in the 1960's, with the advent of single purified

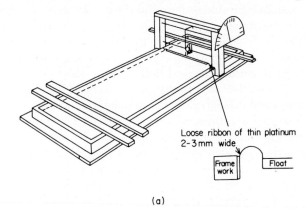

Loose ribbon of thin platinum
2-3 mm wide

(a)

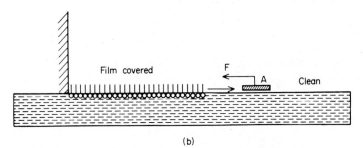

(b)

Fig. 4. (a) Illustration of the principle parts of the Langmuir trough, equipped with torsion wire balance. (b) The packing of a solid film at the air-water interface, between fixed and moving barriers. From Adam, "The Physics and Chemistry of Surfaces." O.U.P., 1941.

phospholipids—purified not only according to class but according to the chemistry of the hydrocarbon chains.[10] The packing of phospholipids in monomolecular films at the air-water interface is controlled primarily by the composition of the hydrocarbon components. At the extremes we have the condensed and the liquid expanded films. The condensed film, which has the characteristics of a solid in the plane of the surface, exhibits a steep force-area curve, indicating a perpendicular orientation of the lipid molecules above the water at all surface pressures. Solid films are characteristic of fully saturated phospholipids, such as diC 18:0 phosphatidyl

choline, and cholesterol, both illustrated in Fig. 5. The natural membrane-
forming lipids all pack as liquid expanded films at the air-water interface.
There is lateral movement of individual molecules in the plane of the
surface which can be seen by blowing on to a packed lipid film of the
expanded type, on top of which a little talcum powder has been sprinkled.
The talcum grains are able to flow about with the surface; on a solid film
the grains do not move.

The molecular characteristics of the biological phospholipids, all of
which produce liquid expanded films, are a high proportion of cis-
unsaturated hydrocarbon chains. Fluidity in the plane of the surface is now
a widely recognised feature of biological membranes and this could allow
for aggregation and disaggregation of monomeric loci (enzymes, receptors,
transport sites, etc.) embedded in the surface. The presence of double
bonds in the cis configuration introduces an irregularity into the simple
dimension of the saturated hydrocarbon chain, which is controlled by the
repeating $108°28'$ tetrahedral bond angle between carbon atoms. The
attractive forces, leading to crystalline array, are optimised by homogeneity
of molecular structure and, as can be seen in Fig 5 (c), the introduction of
only a single cis double bond drastically disturbs the perfection of interaction between neighbouring hydrocarbon chains. This, generally, is the
situation which pertains in biological membranes.

The stability of the surface is the resultant of opposing forces. There is
cohesiveness due to van der Waals interactions between the hydrocarbon
chains and there are electrostatic interactions between the polar head
groups. Of these two factors, the hydrophobic van der Waals interactions
are dominant. The closest stable packing of phosphatidyl ethanolamine
and phosphatidic acid, about 36 $Å^2$ per molecule, is in keeping with the

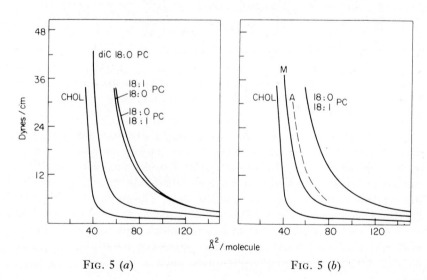

FIG. 5 (a)   FIG. 5 (b)

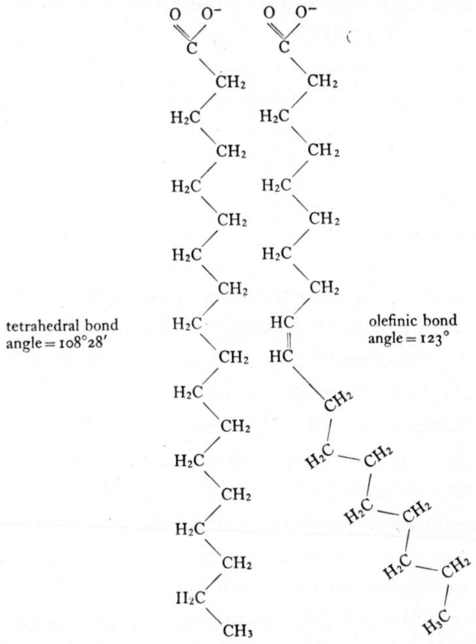

FIG. 5 (c)

FIG. 5. (a) Force area curves of a fully saturated phospholipid (diC 18:0 PC), two mono-unsaturated phospholipids (18:0/18:1 PC and 18:1/18:0 PC) and cholesterol. The ordinate expresses the pressure exerted by the lipid films against the measuring boom, and the abscissa expresses the contained area of a known quantity of lipid floating on the water surface in a Langmuir trough. (From ref. 10.) (b) Force area curves of cholesterol, a mono-unsaturated phospholipid and a mixed film (M) of both compounds in equimolar amounts. The broken line (A) indicates the position of the ideal force area curve for a mixed film obeying the additivity rule. The leftward shift of M from A indicates the extent of film compression in the real situation. (From ref. 10.) (c) The structures of two fatty acids. Stearic acid (C 16:0) is fully saturated. Neighbouring molecules are able to align themselves against each other throughout the length of the chain, and to optimise the van der Waals forces of attraction between them. Oleic acid (C 18:1) has a single double bond in the cis configuration, half way along the length of the chain. The two molecules are drawn so that the "head" ends to C_9 are aligned. Beyond C_9 it is impossible to obtain ideal packing of these two molecules. Even without flexing, the rotation of oleic acid describes the form of a cone (a spinning "Y"), occupying much more space than the simple cylinder of rotation described by the saturated substance. This is the reason why the unsaturated materials (including the phospholipids) display long-range repulsive interactions at the air-water interface, and produce force-area curves of gradual slope. The saturated materials have no long-range interaction, but pack immediately they are pressed into contact to form a solid film; they display a steep force-area curve.

molecular packing of two fully saturated fatty acids. The closest stable packing of phosphatidyl serine and phosphatidyl choline, at about 39 Å² per molecule may be controlled in part by the larger polar head groups. The charge on the head groups (negative for PA and PS; neutral for PE and PC) does not seem to influence the packing properties of these substances.

EFFECTS OF CHOLESTEROL

Sterols are present in roughly equimolar proportion with phospholipid in the plasma membranes of most eukaryotic organisms; they are absent from the membranes of prokaryotes (bacteria). In animals, the main membrane sterol is cholesterol, but other sterols of closely related structure are found in the cell membranes of plants. We should now ask, what are the effects of cholesterol on the surface film properties of phospholipids? This forms part of a very general question, namely, what is the role of sterols in cell membranes? The detailed answers to this question turn out to be complex, and we shall discuss it at other stages in this account. It is one which has been examined by a great variety of biological and physical techniques, on all the currently used membrane models as well as a number of membranes of biological origin. In this and later chapters we shall return to the question of the role of cholesterol, as the various models, membrane species and techniques are introduced in order to try and relate some membrane functional properties with the detailed structure of component molecular units.

We might expect that a monomolecular film formed from a mixture of two surface-forming substances would have an area equal to the sum of the areas formed from the two individual components maintained at the same film pressure. This "ideal" behaviour (which has been termed the additivity rule, and is similar to the familiar Dalton law of partial pressures for mixtures of ideal gases) is not generally adhered to for mixtures of phospholipids and cholesterol, in which an effective compression occurs. The area of the mixed film is invariably less than the sum of the areas of the separate components spread at the same pressure, and this raises the question of whether there is any form of molecular association between the phospholipids and cholesterol?

If cholesterol were merely acting as a space filler in expanded phospholipid aggregates, then it might be expected that steroid molecules of smaller dimensions should equally well produce film compression; but a very thorough study[11] of the film compression phenomenon in mixtures of 18:1/18:0 phosphatidyl choline, with various steroids (see Fig. 5) showed a fairly precise requirement for (1) a planar sterol nucleus, (2) a 3β-hydroxy group and (3) an intact side chain. Thus cholesterol, cholestanol, lanosterol,

7-dehydrocholesterol and B-norcholesterol all interact in 1:1 ratio with 18:1/18:0 phosphatidyl choline so that the molecular area of the sterol-phospholipid pair at 12 dyne cm^{-1} and 37° is about 100 Å². The limiting area of the sterol under the same conditions is about 40 Å² and of the phospholipid, 82 Å², which would give an "ideal" area for the mixed film of about 120 Å². Such a gross reduction in the mean molecular area is probably explained by extensive van der Waals interaction. On the other hand, the smaller molecule, androstan-3β-ol, which lacks the side chain at C_{17} produces no condensation effect. This example demonstrates that film condensation by sterols cannot readily be explained by a non-specific cavity filling phenomenon. The reduced effect of the 3-ketosterols in producing film compression compared to the 3-OH sterols, suggests that the interaction is also stabilised by hydrogen bonding between the OH group and the polar region of the hydrated phosphatidyl choline head group.

The monomolecular film is the most elementary and at the same time most accessible and precise of our membrane models. On the other hand, it is only half of a model membrane and, as such, cannot be used as a model of membrane function. We now move on to consider some "two sided" models of membranes, which are based upon a double molecular layer as their structural element.

PHASE BEHAVIOUR OF PHOSPHOLIPIDS: EFFECTS OF HYDRATION

The presence of two long hydrocarbon chains per molecule of phospholipid might be expected to render these substances extremely insoluble. Unlike the simple soaps (also amphipathic substances), a suspension of phospholipids does not froth on shaking[12]—a certain indicator that the concentration of free monomeric phospholipid is extremely low. Phospholipids even at very low concentrations (critical micelle concentrations $< 10^{-5}$ M), have a natural tendency to aggregate in the presence of water into a number of structurally defined phases.

In the presence of less than 30% aqueous phase, the phospholipids aggregate as extended hexagonal structures[13, 14] in two dimensional array. The dimensions and structural details of the hexagonal phase have been considered mainly by analogy with the soaps, for which much accurate X-ray diffraction data exists.[15] Two general hexagonal phases have been described. The type I hexagonal phase is formed of extended fibres formed with the head groups lining the fibre surfaces. Of the biological phospholipids, only lysophosphatidyl choline forms stable structures of

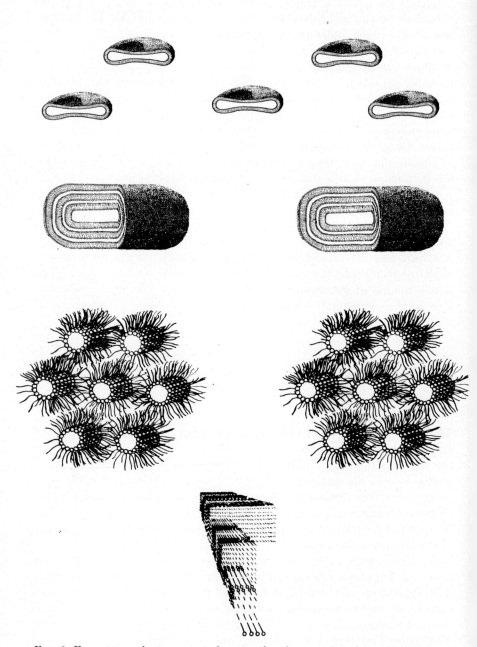

FIG. 6. From top to bottom: monolayer at the air-water interface; single bilayer liposomes (i.e. sonicated liposomes); multibilayer liposomes (the smectic mesophase); hexagonal mesophase, type II and anhydrous liquid crystal. (From ref. 1.) Reproduced by permission of Bangham (1968) from "Progress in Biophysics and Molecular Biology", ed. Butler and Noble, Pergamon Press.

this type. The other phospholipids, at low concentrations of water take up hexagonal phase II. These are water filled, extended, tubular structures (the tubes in hexagonal array), with the internal surface of the tubes lined with the polar head groups.

The biological relevance (if any) of the hexagonal phase is uncertain.

LIPOSOMES

Of much greater importance in biology is the lamellar, or smectic meso-phase, which is stable at high (i.e. natural) concentrations of water. To prepare a lamellar phase suspension,[16] a solution of phospholipid is dried to a thin film on the inner surface of a small glass container, under a stream of nitrogen; aqueous solution is then added, and after shaking hard, most of the lipid will have gone into suspension. This preparation of phospholipids in aqueous suspension is very widely used, and is known as "liposomes".

Much can be learned about the structure of the liposomes by electron microscopy. Because the model membranes do not contain protein, and therefore cannot be "fixed" (Chapter 2, p. 56) by treatment with bifunctional crosslinking reagents such as glutaraldehyde, electron microscopy is generally carried out with the use of negative stain proceedures.[17, 18] By this technique, the stain (ammonium molybdate is preferred to the more common phosphotungstate) penetrates all the water-accessible spaces of the preparation. To reveal the structure of the liposomes, the phospholipid must be hydrated with a solution of the stain. On drying, the stain sets to an electron-dense glass, while the inaccessible regions are revealed as zones of high electron transmission. In this way, liposomes are shown to be particles composed of multiple concentric lamellae[19] ("onion skins"). In the negative stain presentation, each electron translucent layer is the image of a phospholipid bilayer. The polar headgroups are turned outwards from the layer into aqueous zones which are electron dense due to the presence of the stain. [1, 19]

This preparation, the liposomes, are the first of our membrane models having functional properties. It is possible to form liposomes with one aqueous solution (e.g. radioactive KCl) and then wash them free of external radioactive isotope by simple centrifugation. Thus, it is possible to follow transport kinetics in situations where no electrochemical gradient exists (i.e. by equating the chemistry of the internal and external aqueous solutions), by measuring the rate of appearance of the label in the external solution.[12] Alternatively, we can make use of the fact that the permeability of the liposomes to water is so much greater than to most solutes that by setting the solute concentration on the outside different to that on the inside, the liposomes will swell or shrink rapidly as a result of the osmotic movement of water. The subsequent slow movement of solute can then be

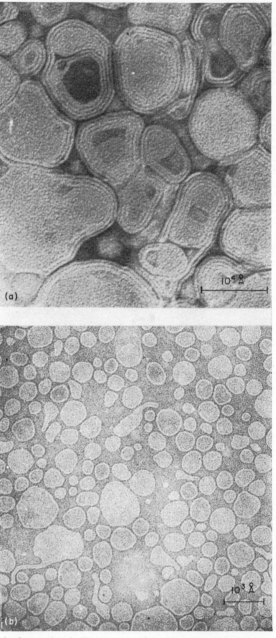

F<small>IG</small>. 7. Negatively stained liposomes. (*a*) Multilamellar liposomes (smectic mesophase) prepared from a mixture of egg yolk lecithin (96%) and phosphatidic acid (4%) and negatively stained with ammonium molybdate. (From ref. 18.) (*b*) Sonicated single bilayer liposomes. (From ref. 23.) Reproduced by permission of Bangham (1974), *Methods in Membrane Biology*, **1**, 1.

followed by measuring the change in light scattering in a spectrophoto-meter, as the liposomes gradually relax to their original dimensions.[20]

With liposomes, it is possible to relate their kinetic-functional properties to their composition. Also, because of their size and light scattering properties, they can be seen in the microscope at low power, using dark field optics. Individual observation of liposomes which have been produced by shaking up phospholipids in aqueous media is used in the investigation of the electrostatic properties of phospholipid surfaces by electrophoresis in voltage gradients.[21, 22] In these ways, liposomes can be treated as if they were cells: the same functional questions can be asked, and the same techniques can be used.

Single Compartment, Single Bilayer Liposomes

A closely related phospholipid membrane model is the sonicated liposome. A suspension of phospholipid liposomes is prepared by shaking as before. They are then subjected to sonic irradiation for as long as is needed to clarify the suspension. After sonication, a little heavy material sometimes remains: this is removed by ultracentrifugation. The supernatant fluid should be quite clear with a slight bluish tinge. Electron microscopic examination of this preparation (sonicated in the presence of ammonium molybdate as a negative stain) reveals that the buoyant material is composed of vesicles having a single limiting membrane.[23] As with the shaken lipo-some preparation described previously, the sonicated liposomes may be prepared with separately defined aqueous phases, inside and out. They are a little more difficult to handle, primarily because they are buoyant, and so cannot be sedimented by centrifugation. In seeking to exchange the exterior medium we can use the techniques of gel filtration or dialysis.

Due to the maximal exposure of the polar head groups with the external aqueous environment in this preparation, it is possible to study the inter-action of water-soluble substances (e.g. electrolytes, polypeptide hormones, drugs, immunoglobulins, etc.) with phospholipid surfaces of defined composition using spectral methods (e.g. NMR). Also, the sonicated liposome preparation is widely used in the reconstitution of lipid-dependent enzymes which have to be purified in an inactive "delipidated" condition. But as models for kinetic studies of solute transport they are really not so useful as multilamellar liposomes, although one might have thought that their single compartment structure would have made them more "cellular". The problem lies in their size. A very careful investigation[23] of the dimen-sions of sonicated liposomes formed from a mixture of phosphatidyl choline (96%) and phosphatidic acid in 0·16 M KCl has been made. These are single compartment, single bilayer shells of external radius 120 ± 4 Å, and internal radius 73–96 Å. The volume per μmole of lipid is $3·08.10^{-4}$ cm^3 (assuming a spherical vesicle), and the external surface area per μmole of lipid is 2830 cm^2. The mean aqueous "core" volume of each liposome is

7.10^5 Å^3. Each vesicle thus contains about 37,500 water molecules (with water at 55 moles/litre) and if filled with physiological salt solution, would contain about 50 each of anions and cations. The concentration of trace labelled ions is bound to be low.

THERMAL PHASE BEHAVIOUR OF PHOSPHOLIPIDS[24, 25]

We consider now the thermal phase behaviour of the phospholipids, starting with single species in anhydrous form. The anhydrous phospholipids are prepared by heating at 90° for four hours: even in a desiccator over P_2O_5 they revert to the monohydrate form. Like other fatty materials, the phospholipids display typical melting behaviour, and change to a free flowing fluid phase at a characteristic capillary melting temperature. This is generally about 100° above the capillary melting temperature of the simple fatty acids which are esterified to glycerol in the phospholipid. When the solid material is observed by polarisation optics while slowly raising the temperature on a microscope equipped with a heating stage, another phase transition may be detected.[24] This occurs at about the normal melting temperature of the free fatty acids (i.e. about 100° below the capillary melting temperature of the phospholipid) and at this point the material changes from a transparent solid to an anisotropic liquid-crystalline phase. These transitions are true phase changes in the thermodynamic sense of this term, and are characteristically accompanied by a flow of heat.

The heat flows at the transition points can be measured by differential scanning calorimetry (DSC).[24, 26] In this technique, heat is supplied to the sample (and to a relevant reference) so that the temperature increases at a constant rate with time. So long as the specific heat of the sample remains constant, the rate of supply of heat to the sample will be constant too, but when the temperature of an endothermic phase change is reached, the heat supply must be increased in order to maintain the linear rate of temperature increase. It is the difference in the energy supply to the sample and to the reference (e.g. an empty sample container or, for solutions, an identical quantity of the solvent) which is actually recorded, and the optimum temperature for the phase change is the temperature at which the trace first leaves the base line: the peaks can be integrated to provide a measure of the energy absorbed or yielded by the sample at the phase transition.

A series of differential scanning calorimetry curves for diC 18:0 phosphatidyl choline in the presence of increasing amounts of water is illustrated in Fig. 8.[27] The topmost curve is for the anhydrous material, and there is a single, rather broad thermal transition at 77°. As water is added to the system, the transition temperature declines to 57° and the DSC scan sharpens, suggesting that the water produces a loosening of the structure of the phospholipid liquid crystal and allows the molecular aggregate to melt cooperatively, all in one piece. No independent phase transition at 0°

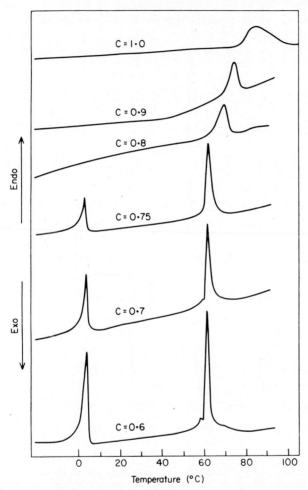

FIG. 8. (a) Differential scanning calorimetry heating curves for diC 18:0 PC in the presence of increasing amounts of water. The peaks in the traces indicate the input of heat at the phase change. The peak at 0° is due to the melting of ice; the peak at and above 61° is due to the melting of the phospholipid hydrocarbon chains. (From ref. 27.)

due to the melting of water is detectable until the weight fraction of water exceeds 20%, indicating that each molecule of phospholipid may bind up to 10 molecules of water to form a single hydrated lipid phase. In a biological membrane the bound water should be regarded as an intrinsic part of the membrane structure and organisation.

The effect of varying chain length in a series of symmetrical fully saturated phosphatidyl cholines, dispersed as liposomes in an excess of water, is shown in Fig. 8.[26] Here it can be seen that as the chain lengths are reduced from C_{18} to C_{14}, the temperature of the main phase transition drops from 57° to 27°, paralleling the melting points of the related fatty

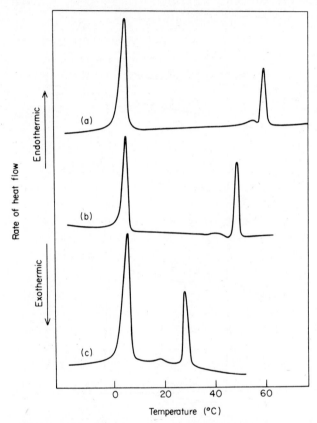

FIG. 8. (b) Differential scanning calorimeter heating curves for aqueous suspensions of fully saturated phosphatidyl cholines (a) diC 18:0; (b) diC 16:0; (c) diC 14:0. (From ref. 26.)

acids which decrease with shorter chain lengths. The *capillary* melting temperatures of the phospholipids (PE 200°, PC 230°) are quite independent of the hydrocarbon components and depend entirely on the nature of the head groups. The presence of double bonds in the hydrocarbon chains also depresses the temperature of the intermediate phase transition compared with the corresponding fully saturated phospholipid and this is particularly true of phospholipids containing cis-double bonds, which are characteristic of biological material.

The spectral properties of phospholipids as single phases and as aqueous dispersions (liposomes) also alter at the thermal phase transitions. Both infra-red (IR)[24, 28] and nuclear magnetic resonance (NMR)[24, 29] spectroscopy give a picture of gradually increasing movement of the hydrocarbon chains as the phase transition is approached from low temperatures. At the transition point, there is a fundamental change: in the IR spectrum, all fine structure disappears, and it now resembles the spectrum of a solution of the phospholipid in chloroform. The NMR spectrum sharpens.

The low angle X-ray diffraction pattern of the anhydrous material (indicative of long spacings in the repetitious structure (see Chapter 2, p. 59)) becomes diffuse, and this indicates a change to a less ordered arrangement of the hydrocarbon chains. By contrast, the wide-angle diffraction pattern (indicative of short spacings) changes little, and this shows that the space occupied by the polar head groups remains essentially constant.[24] The picture is one of a sudden increase in fluidity in the hydrocarbon phase. Above the phase transition there is (within the two-dimensional plane) a highly permissive freedom of movement of phospholipid molecules. Immediately below the transition point, all movements, both inter- and intramolecular, become acutely restricted, although it cannot be said that the hydrocarbon is truly frozen; as the temperature is lowered beyond the transition point, the limitations on motional freedom continue to be imposed still more stringently right down to the temperature of liquid nitrogen.

This is an appropriate point to consider briefly some of the reasons underlying the characteristically differing behaviour of the fully saturated phospholipids and their (cis) unsaturated counterparts which are generally found in biological membranes. Two features stand out. (1) Saturated phospholipids pack as solid films, occupying minimal area per molecule, even at low film pressures at air-water interfaces: unsaturated phospholipid molecules interact with each other over much greater distances, and much greater forces are required to pack them to their limiting minimal areas. (2) The temperature of the thermal phase transition is lower for unsaturated than for saturated phospholipids. In considering the stability of phospholipid lamellae we are primarily interested in the tendency of neighbouring hydrocarbon chains to pack alongside each other, maximising the van der Waals interaction between them. Such interactions arise from the temporary distortions in the time-average symmetrical fields about these uncharged molecules (sometimes called London dispersion forces), which diminish according to the seventh power of the distance, and are therefore appropriately termed "very short range forces". Quite clearly, the more regular structures arising from the packing of the fully saturated phospholipids which allow maximum contact throughout the length of the fully extended hydrocarbon chains will produce more stable aggregates. The longer the chains, the larger is the surface of contact between neigh-

bours—and the higher the stability, as is demonstrated by a higher
transition temperature. The presence of a bend or a kink (cis double bond)
in the chain reduces the possibility of contact between neighbours; the
stability of the aggregate is less, and the temperature of the thermal phase
transition is lower. An unsaturated phospholipid molecule allowed rota-

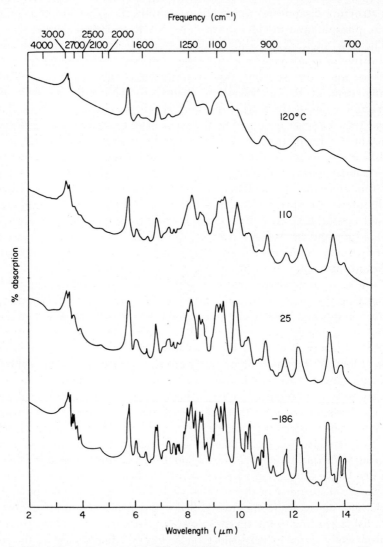

FIG. 9. The infra-red spectra of diC 12:0 PE at different temperatures. The fine
structure of the spectrum is gradually reduced as the temperature is raised from
$-186°$ to $110°$, indicating a gradual loosening of the hydrocarbon chains in the gel
phase lipid. The sudden loss of spectral detail between $110°$ and $120°$ is due to the
transition into the liquid crystalline phase. (From ref. 28.)

tional freedom along the axis of one of its hydrocarbon chains, may be visualised as having the shape of a spinning Y. It occupies more space than its fully saturated counterpart, and will begin to interact with its neighbours over much greater distances. The forces of attraction between the molecules due to van der Waals overlap are less, due to the structural irregularities and the consequently reduced area of contact between neighbours; much greater forces have to be applied in order to maximise the packing density.

In turning one's thoughts towards biological membrane structures, it is justifiable to ask whether the consideration of single phospholipids (be they saturated or unsaturated) is directly relevant. This question may be answered affirmatively, but only in a limited sense. There are prokaryotic organisms in which it is possible to manipulate, through the composition of the growth medium, both the sterols and the hydrocarbon components of the membrane phospholipids. It is thus possible with cultures of *Acholeplasma laidlawii* to generate authentic biological membranes having no cholesterol, and only a unique phospholipid hydrocarbon. In almost all other situations we are of necessity concerned with phospholipid mixtures and, in general, with membranes deriving from the cells of eukaryotes, we are concerned with admixtures of phospholipids with sterols as well.

When we compare the thermal phase behaviour of mixtures of phospholipids with single substances, we see that the temperature range over which the phase transition occurs becomes broadened.[25, 30] This is well illustrated by differential scanning calorimetry. Figure 10 (*a*) shows DSC traces for a series of binary mixtures containing diC 14:0 PE and diC 14:0 PC in various proportions, and dispersed as liposomes in water (lipid:water 50:50). Whilst the temperature ranges over which the pure compounds absorb heat are narrow, the scans for the binary mixtures are broadened, and highly asymmetrical. Such broadening is not seen for binary mixtures of lipids sharing a common head group, but differing in the chain length of the acyl hydrocarbons. Broadened transitions, which are also characteristic of some biological membranes (e.g. those of *Acholeplasma laidlawii* and micro-organisms lacking cholesterol) most probably arise from the simultaneous presence of distinct fluid and frozen clusters of phospholipid molecules in the bilayer; a phenomenon referred to as "phase separation".

EFFECTS OF CHOLESTEROL

(a) *Calorimetric Measurements*
Earlier, we considered the question of phospholipid-cholesterol interaction from the point of view of molecular packing and compression on monolayer films (p. 12). Cholesterol also affects the thermal phase properties of phospholipids and this can be investigated by calorimetric and spectral methods. A series of differential scanning calorimetry curves of diC 16:0

B

PC-cholesterol mixtures dispersed as liposomes in water are shown in Fig. 10 (*b*).[38] The topmost curve is for the pure phospholipid, and the absorption of heat at 41° due to phase change is clearly marked, together with a minor transition at 33°. As the fraction of cholesterol in the mixture

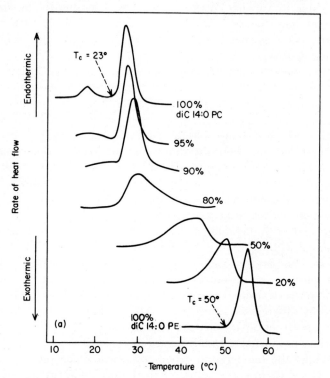

FIG. 10. (*a*) Differential scanning calorimeter heating curves of a two component mixture containing diC 14:0 PC and diC 14:0 PE, in the molar proportions indicated on the right-hand side of the figure. Reproduced by permission of Chapman (1974), *J. Biol. Chem.*, **249**, 2512.

is increased to 12·5%, the absorption of heat due to the minor transition disappears, and this is followed by the loss of the major phase transition as the fraction of cholesterol is raised from 20% to 50%. The effect of cholesterol is to disrupt the highly ordered array of the saturated hydrocarbon chains, so that when the cholesterol and the phosphatidyl choline are present in equimolar proportions, all the hydrocarbon is in a fluid condition, even at temperatures below the normal freezing point, and there is a minimum of interaction between like molecules.

(b) *Nuclear Magnetic Resonance Measurements*
The fluidity of the natural phospholipids at ordinary temperatures is such that sharp high resolution proton magnetic resonance spectra (NMR) can

be obtained. In this technique, the spectral lines can be assigned to individual groupings in the material under study. A variety of factors, both intra- and extramolecular, affect the local magnetic field experienced by any individual proton nucleus, and thus the applied field strength at which the nucleus resonates. These factors control the degree of chemical shift of the proton resonance in the spectrum, compared with that of a

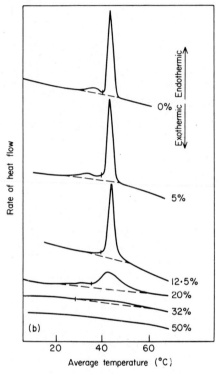

FIG. 10. (*b*) Differential scanning calorimeter heating curves of a two component mixture containing diC 16:0 PC and cholesterol, in the molar proportions indicated on the right hand side of the figure. Reproduced by permission of Ladbrooke *et al.* (1968), *BBA*, **150**, 143.

reference (commonly the protons of the chemically equivalent methyl groups of tetramethyl silane, or a suitable water-soluble reference compound). High resolution NMR spectra show separate signals associated with different functional groups in a molecule, as can be seen clearly in the NMR spectrum of egg yolk lecithin dissolved in $CDCl_3$,[32] illustrated in Fig. 11 (*a*). (*N.B.* Deuterated solvents are used in order to avoid the overwhelming obscuring effect that solvent protons would have on the spectrum: chemical shifts are measured in parts per million; the resonance of deuterium is removed by many MHz from that of protons.) The areas lying under the peaks in the spectrum are directly related to the relative

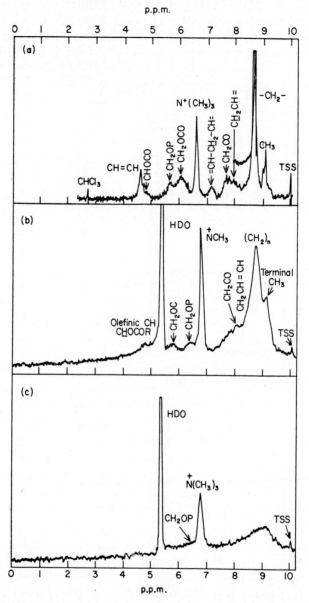

FIG. 11. Nuclear magnetic resonance spectra of egg yolk lecithin. (a) Solution in $CDCl_3$; (b) dispersed in D_2O; (c) dispersed as an equimolar mixture with cholesterol in D_2O. The small sharp peak at 10 ppm is due to the unshifted proton resonance reference. The large —CH_2— peak in (a) has been recorded a second time at reduced sensitivity. (a) from ref. 32; (b) and (c) reproduced by permission of Chapman and Penkett (1966), *Nature*, **211**, 1304.

contributions of individual protons in the molecule; thus the height of the methyl resonance (CH_3—; six protons per molecule) is about two-thirds of the height of the N-methyl resonance (($CH_3)_3N^+$—; nine protons per molecule) with both spectral peaks having an approximately similar band width. The area, and hence the peak height of the —CH_2— protons is so great that it has been recorded at reduced sensitivity.

Figure 11 (b) shows the NMR spectrum of the same material (egg yolk lecithin) dispersed as liposomes in D_2O.[33] On comparing the dominant lines of this spectrum with their assigned counterparts in Fig. 11 (a), it will be seen that some have shifted their positions and others are broadened. Most noticeably, the resonances due to the protons of the hydrocarbon component of the phospholipid (CH_3—, —CH_2—, —$CH=CH$—, etc.) have broadened spectral peaks, but these remain in the same positions as in the solution spectrum; the head group protons ($N^+(CH_3)_3$–, CH_2OC, CH_2OP) have shifted spectral peaks, and only the $N^+(CH_3)_3$– protons give rise to a sharp peak. Line broadening occurs when the molecular motion of a particular group becomes restricted either in amplitude or frequency, and this is what has happened to the protons lying in the hydrophobic layers of the liposomes. The motion of the terminal choline head group is unrestricted, but its shifted position in the NMR spectrum (and the peaks due to CH_2OC and CH_2OP) indicate that these protons are sensitive to the change in environment in going from solution in $CDCl_3$ to suspension in water. The shifted protons are those which "see" the water; the unshifted protons are those which are buried within the hydrocarbon domains of the liposome structure, an environment which closely resembles that of the organic solvent of Fig. 11 (a) but which now imposes some restriction of movement due to its intrinsic viscosity.

Figure 11 (c) illustrates the NMR spectrum of an equimolar mixture of egg yolk lecithin and cholesterol, dispersed as liposomes in D_2O.[33] All fine structure in the spectrum has now disappeared due to line broadening, except for the line due to the choline protons. Most strikingly, there is no spectral expression of any protons due to the cholesterol; this is due to the rigid plate-like structure of this molecule, which allows very little freedom of movement of its constituent parts. Cholesterol prevents the movement of hydrocarbon chains which would normally be in the fluid condition (Fig. 11 (b)) above the phase transition temperature: the choline protons, lying well away from the cholesterol molecule remain unhindered, and their sharp line spectrum is retained.

By comparison, with phospholipid systems which would normally be below the thermal phase transition temperature, the effect of adding cholesterol is to enhance hydrocarbon chain movement, and the NMR spectra of gel phase phospholipids are sharpened in its presence. Cholesterol causes phospholipids of differing hydrocarbon chain composition to take up an "intermediate fluid phase". This buffering effect of cholesterol on the thermal phase properties of hydrated phospholipids seems to be true too of the effect of cholesterol on biological membranes. Myelin and red cell membranes both have a high proportion of cholesterol, and no phase transition is apparent from spectral or calorimetric examination of either the intact membranes or lipid extracts prepared from them.[27, 34] In contrast, biological membranes having a low proportion of cholesterol (such as plasma membranes from bacteria and mitochondrial inner membranes) generally do show phase transitions.[35, 36, 37] These are broad and only partly cooperative, such as characterise the phase transitions of mixtures of phospholipids having hydrocarbon chains of differing chain length and unsaturation (see p. 23.)

The concept of fluidity in the plane of the bilayer has become one of the central tenets of modern membrane biology and, certainly, a limited state of controlled lateral fluidity in the bilayer of membranes is probably necessary for the support of life.[38] The thermal characteristics of a number of activities associated with membranes, such as transport,[39] assembly of membrane proteins,[40] and other enzymes[41, 42] all show abrupt changes in rate at temperatures which may be related to the phase behaviour of the membrane lipids. The membranes of non-homothermic organisms (cold-blooded animals, plants and micro-organisms) have to be capable of functioning over wide ranges of temperatures, and the membranes of bacteria, which lack the buffering contribution of cholesterol may be subject to changes of state which might have drastic effects on their structural and barrier properties. It has often been remarked that bacteria cultivated under conditions of cold contain increased proportions of unsaturated and short chain phospholipids in order to maintain the membrane in a fluid condition[43, 44]. It is likely that the membranes of eukaryotic heterotherms are also carefully regulated in order to ensure that a condition of fluidity is maintained in the face of wide fluctuations in ambient temperatures and availability of nutrients, though, of course, these higher organisms would be protected to some extent due to the presence of cholesterol (or a suitable analogue).

In considering the concept of fluidity in membranes, it must be stressed that it refers only to the movement of molecules laterally in the plane of the membrane. The transfer of a phospholipid across the membrane to a position in the opposing monolayer ("flip-flop") would be a most improbable event. In this respect, the term "fluidity" as applied to membranes does not have the same meaning as when it is applied to bulk phases. The reason why the movement is restricted in two dimensions lies in the amphipathic nature of the component phospholipids and the intrinsic proteins found in membranes, and is due to the very favourable interaction between the polar head groups and the external aqueous environment and the very great difficulty of transferring polar groups into the low dielectric medium of the interior (see p. 45).

LIPOSOMES: "FUNCTIONAL PROPERTIES"

When liposomes which have been prepared by shaking phospholipid in a salt solution of one concentration, are resuspended in a medium of different concentration, they swell or shrink, depending on the osmotic difference of the enclosed material and the new suspending solution. The swelling and shrinking of liposomes, which arises from the movement of water tending to equalise the osmotic pressure on either side of a semi-permeable barrier, forms the basis of a method for measuring solute fluxes.[20] Changes in the optical density, which arise from changes in the light scattering properties of the suspension are inversely proportional to the volume enclosed by the liposomes, i.e. $\dfrac{dV}{dt} = \dfrac{d1/A}{dt}$. Measurements are made with a spectrophotometer, at an "inert" wavelength, generally around 450 nm.

In Fig. 12 the faint lines show how KCl loaded liposomes respond when transferred to a number of different iso-osmotic solutions.[20] Increased swelling is indicated by a downward movement. The heavy line represents the volume increase as the KCl loaded liposomes are transferred to water. Transfer of KCl loaded-liposomes to iso-osmotic solutions of KCl, NaCl, Na acetate, glucose, sucrose and mannitol produces no essential change in the mean volume of the liposomes. On the other hand, transfer to water and iso-osmotic solutions of urea, glycerol and ammonium acetate produces a decrease in light scattering due to penetration of the liposomes by these substances and consequent swelling. Ethylene glycol, methyl urea and ethyl urea penetrate at least as fast as water; propionamide and glycerol penetrate less fast and malonamide and erythritol are much slower. The order of permeability of non-electrolytes resembles that for red blood cells, and so this observation may be said to represent the first indication that in working with phospholipid liposomes we are studying a model that resembles a biological membrane in at least one important functional respect. The high permeability to ammonium acetate is worthy of comment, especially in view of the fact that liposomes are impermeable to sodium acetate. This probably arises from the weak acid and base properties of the acetate and ammonium ions, the real penetrating forms being ammonia and acetic acid. Osmotic swelling is quite reversible, as may be seen when non-penetrating KCl is added to make the aqueous medium iso-osmotic with the initial liposome contents after swelling has taken place.[20] With regard to the movement of electrolytes, the liposome preparations do not appear to mimic biological membranes well. In order to detect movement of electrolytes, radioactive isotopes are used. Typical $^{42}K^+$ fluxes are of the order of $0 \cdot 5.10^{-15}$ equiv. cm^{-2} sec^{-1} (equivalent to a D.C. resistance of $10^{10}\ \Omega$ cm^2) and this is many orders of magnitude lower than the K$^+$ fluxes through most cell membranes.[12, 45]

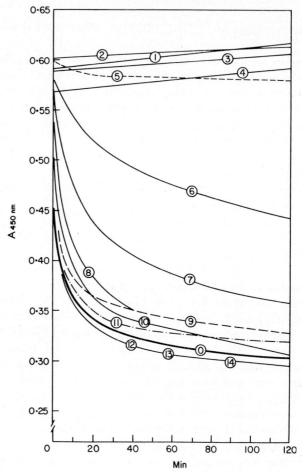

Fig. 12. The changes in light scattering by multilamellar liposomes suspended in solutions of penetrating and non-penetrating solutes. The liposomes were hydrated in a solution of KCl, and then small amounts were added to cuvettes containing either distilled water (heavy line) or iso-osmotic solutions of (1) glucose; (2) KCl; (3) NaCl; (4) sucrose; (5) sodium acetate; (6) erythritol; (7) malonamide; (8) urea; (9) propionamide; (10) glycerol; (11) ammonium acetate; (12) methylurea; (13) ethylurea; (14) ethylene glycol. A downward sweep indicates a reduction of light scattering (decreased optical density) due to inward movements of permeant solutes and subsequent movement of water. (From ref. 20.)

EFFECTS OF CHOLESTEROL

We have considered the effect of cholesterol on the monolayer film packing characteristics and on the thermal phase transitions of the phospholipids: at a functional level, the role of cholesterol in combination with phospholipids can be investigated by making measurements of solute fluxes across liposomes of defined composition.

The dual effect of cholesterol on phospholipids is illustrated by its effect on the permeability of liposomes to non-electrolytes.[46] The permeability of liposomes to glycol and glycerol was measured at different temperatures, using the light-scattering technique described above (see Fig. 13). Egg lecithin and three saturated phospholipid preparations were used, each one with and without cholesterol (30 moles %). In all cases, the permeability to the smaller glycol molecule is greater than the permeability

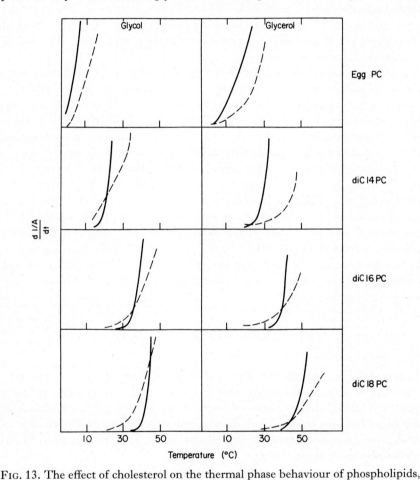

FIG. 13. The effect of cholesterol on the thermal phase behaviour of phospholipids, revealed by its effect on the permeability of liposomes to non-electrolytes. The permeabilities are estimated from light scattering measurements in which the initial swelling rates are determined from the relationship $\dfrac{dV}{dt} = \dfrac{d1/A}{dt}$. The solid lines indicate the temperature dependence of the permeability of pure phospholipids (defined phosphatidyl cholines plus 4% of phosphatidic acid). The broken lines indicate the temperature dependence for systems containing 30 moles % of cholesterol. Reproduced by permission of de Gier et al. (1969). BBA, **173**, 143.

to glycerol. The effect of cholesterol on the swelling rates of all the preparations is to reduce the temperature dependence: the sharp thermal transitions are lost (see for an extreme example the glycol permeability change at 40° for diC 18:0 phosphatidyl choline). At low temperatures the permeability of the fully saturated synthetic phospholipids in the presence of cholesterol is enhanced; at high temperatures the permeability is reduced. Turning now to a natural phospholipid (egg lecithin) we recognise that the hydrocarbon portion is (of course) much more mobile than the saturated synthetic substances at lower temperatures. Here the effect of cholesterol is only to reduce permeability: there is still some reduction in the temperature dependence.

Different steroid substances have been tested in combination with phospholipids to determine the structural requirements for an effect on liposome permeability.[47] Out of a large number of sterols tested in combination with egg lecithin (see Fig. 3), only cholesterol, cholestanol, lanosterol, 7-dehydrocholesterol and B-norcholesterol showed a pronounced effect in reducing the permeability to Rb^+, glucose and glycerol.

The same criteria seem to apply in the regulation of liposome permeability as apply in the compression of monomolecular films at air-water interfaces (see p. 12); that is to say, there is a requirement for (1) a planar sterol nucleus, (2) a 3β-hydroxy group and (3) an intact side chain. Thus, coprostanol (cis-structured at the A/B ring fusion, and therefore non-planar), androstan-3β-ol (no side chain) epicholesterol and androstan-3 α-ol (3α-hydroxysterols) are all without effect on the permeability of egg lecithin liposomes. Some of the ketosteroids even increase permeability.

BILAYERS: AIR INTERFACES

In a number of respects the liposome as a model of biological membrane structure is limited. Primarily, this is because it is not possible to make electrical measurements. Only one side of the liposome is accessible after the phospholipids have been hydrated, and so it is not possible to alter or probe the interior contents. The planar black lipid membrane (BLM) allows one to do these things.

The presence of black or non-reflective spots on soap bubbles is an ancient observation. Newton in 1704 clearly recognised that the black spots are not holes, but exceedingly thin regions in the soap film. He went further than this; he actually made measurements of the thickness of black films, and came up with a result which seems the more astonishing to us today with our knowledge of membrane biology. He recorded his result as "$\frac{3}{8}$, expressed in parts of an inch divided into ten hundred thousand equal parts".[48] We would say 95 Ångstroms.

The transition from a thick coloured film to thin black film in a soap bubble has all the elements of a phase transition. The reflective film is thick: that is to say, it is many orders of wavelengths of light thick. The interference colour pattern, which is repeated each time the thickness of the film changes by one-quarter of a wavelength (green light, $\lambda \sim 5500$ Å) is indicative of this: and the pattern is repeated many times across the film. The black region is exceedingly thin with respect to the wavelength of light, the blackness itself arising from the cancellation of the two emergent reflected rays. Compared to the wavelength of the visible light, the path length difference between the ray reflected from the front face and the back face of the thin film is negligible. As there is a 180° phase change at the front face, the two rays emerge out of phase, and there is cancellation. The film appears black although it is very transparent.

The thickness of the black soap film is just twice the length of the hydrocarbon chains of the soap molecules. As with the monolayer at the air-water interface, the hydrocarbon chains project out into the air, and the polar head groups interact with each other at the centre of the film. The surface of thick (reflecting) films are organised similarly, but the interior is considerably disordered.

BILAYERS: AQUEOUS INTERFACES

Apart from its thickness, there is not much resemblance between a soap bubble at air interfaces and biological membranes. In 1961, Mueller, Rudin, Tien and Westcott first showed how to construct black films from biological membrane phospholipids, separating two aqueous compartments.[49] In their method,[50] a paint brush is used to apply a solution of phospholipids in organic solvent across a 1 mm diameter orifice in a small "Teflon" cup. The cup is partly submerged in aqueous solution which covers the orifice, so that the film of phospholipid applied to the orifice separates the aqueous solution into two compartments. The film can be illuminated, and observed through a simple lens system. As with the soap films at air interfaces, the interference patterns characteristic of thick films are first observed. The film thins as the material drains and the solvent diffuses into the aqueous compartments.[51] The black regions generally appear at first as small spots (Fig. 15) which coalesce until the orifice is mainly covered by a non-reflective thin film. The principles of molecular organisation in bilayer films are illustrated in Fig. 16. That the film is coherent and not broken can be checked by measuring the electrical resistance across it. Black films[3] characteristically have extremely high resistance—up to 10^8 ohms cm^2, and in this respect they do not resemble biological membranes[52], the resistances of which are mainly in the range 10^3-10^5 Ω cm^2.

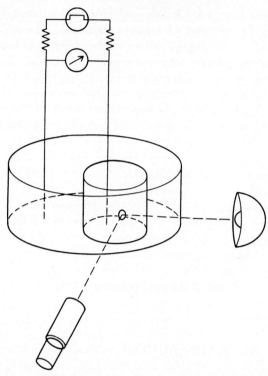

FIG. 14. A simple chamber for measurements on black lipid membranes. A 5 ml Teflon cup rests inside a glass petri dish. Saline solution level is above the level of the hole across which the phospholipid solution is painted with a camel-hair brush. The membrane is observed at $10 \times$ magnification under reflected light. The rudiments of a conventional pulsing and recording circuit are shown. (From ref. 51.)

In total contrast with biological membranes, the black lipid membranes are almost totally non-selective among cations, and further, the electrical conductance (slight though it is) obeys Ohm's law, i.e. the membrane current, I, is proportional to the membrane voltage, V. At a biological level, we look for relatively high conductance; well-developed selectivity between different monovalent cations (100:1 selectivity ratios and above); and in many cases a variable resistance and rectification characteristic, which might be regulated by transmembrane electric or redox potentials, the presence of "transmitter" substances, the effect of light, etc. The basic relationships between voltage and current, for systems obeying and systems diverging from Ohm's law, are illustrated in Fig. 17.

CAPACITANCE AND BILAYER THICKNESS

The capacitance of resistive barriers can be simply estimated from measurements of the time required to charge or discharge the potential, when

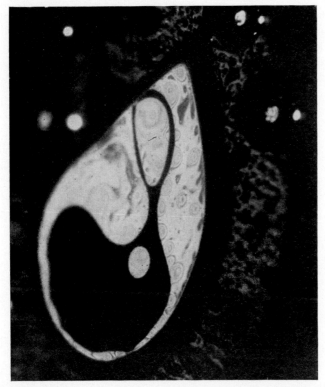

FIG. 15. Photograph of an underwater black lipid membrane in the process of formation. This membrane was formed across a steel wire loop (2 mm diameter) immersed in NaCl (0·1 M). The photograph, taken with reflected light, reveals the thick film which has a thickness at least equal to the wavelength of visible light (5000 Å). The bilayer region (100 Å) is invisible, but is able to provide mechanical support for the thick film. (From ref. 51.)

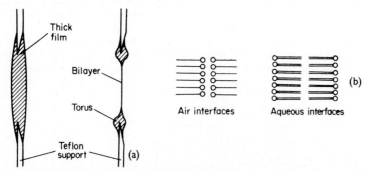

FIG. 16. (a) Diagrammatic representation of bilayer formation from a lipid film across a small hole. Most of the solvent diffuses into the aqueous phase, and the bulk of the lipid drains to form a "torus" of bulk phase lipid solution around the black membrane which develops spontaneously. (b) Schematic representation of the arrangement of amphipathic molecules as bilayers at air and aqueous interfaces.

sudden step changes in the applied voltage are made.[50] For phospholipid bilayers made by the brush technique, the capacitance is generally found to be somewhat lower than that of a number of biological membranes for which the capacitance is known.[6, 53] Capacitance measurements can be used to determine the thickness of the resistive barrier so long as the

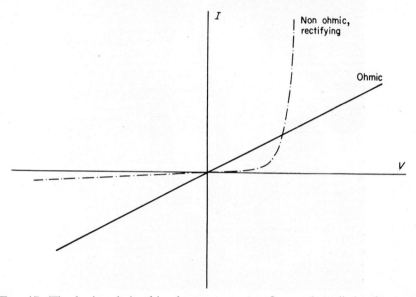

FIG. 17. The basic relationships between current flow and applied voltage are illustrated by an *I-V* curve. A straight-line relationship indicating strict proportionality between *I* and *V* demonstrates adherance to Ohm's law, $V = IR$. When the slope of the line is steep, the resistance is low and the conductance ($= 1/R$) is high. Conversely, a shallow slope is indicative of high resistance and low conductance. The model membranes obey Ohm's law, and have very low conductance.

In biological systems, various non-linear relationships between *I* and *V* are found. The illustration shows a system exhibiting rectification and voltage dependent conductance, i.e. the system only conducts when the voltage is applied in a particular direction, and when it is in the current conducting mode, the conductance (I/R) varies with the applied voltage. This is characteristic of bilayer membranes which have been treated with the antibiotic alamethicin (see Chapter 3, p. 158).

dielectric constant of the resistive material is known. For membranes based on the phospholipid bilayer this is generally assumed to be the same as the dielectric constant of the related paraffin hydrocarbons. By making the relevant substitutions in the equation

$$C = \frac{\epsilon A}{4\pi d}$$

where A = area of membrane
ϵ = dielectric constant of the resistive material
d = thickness of the resistive layer,

it can be shown that the thickness of phospholipid bilayers is somewhat greater than that of the membranes from which the phospholipids are extracted. To this extent the planar phospholipid bilayer is not an ideal model for a biological membrane. Of course, the phospholipid bilayer contains impurities in the form of the solvents in which the phospholipids were originally dissolved. These impurities will be distributed to a variable extent between the bilayer itself, the surrounding "torus" and the bulk aqueous phase, depending on the composition of the materials used. There is certainly a sufficient amount of solvent retained in the bilayer as normally prepared to affect its thickness and capacitance because it is possible to increase the capacitance and reduce the thickness of the

TABLE 1. Influence of hydrocarbon solvents on the capacitance and thickness of bilayer membranes formed from egg yolk lecithin.[6]

Solvent	C (μF/cm^2)	d (Å)
n-decane	0·385	46·5
n-dodecane	0·443	48
n-tetradecane	0·515	40·1
n-hexadecane	0·603	32·4

resistive barrier of the bilayer by using progressively longer chain hydrocarbons as solvents for the phospholipids. The effect of a homologous series of paraffin hydrocarbons on the thickness and capacitance of (egg yolk) PC bilayers is illustrated in Table 1.

The reason for the thinning in the membranes formed with longer chain solvents is the progressive exclusion of these materials from the bilayer, until the point is reached with hexadecane that only negligible quantities remain. One might have thought that PC-hexadecane would be a composition of choice for the making of planar bilayers but there is the problem that membranes formed from this particular mixture tend to expand continuously with time because of the continuous recruitment of material from the surrounding torus into the bilayer, which eventually extends well beyond the limits of the orifice in the teflon support. Best of all would be to form phospholipid bilayers in the absence of any contaminating lipid solvents, and a technique for doing this has been devised.

Solvent-free Bilayers
In this method,[53] lipid monlayers are formed initially at the air-water interface, in two chambers (resembling Langmuir troughs) separated by a very thin hydrophobic "Teflon" film which is pierced by a small orifice. When the level of the water in the two chambers is slowly raised, the monolayers are deposited on the film, and at the site of the orifice a bilayer develops as shown in Fig. 18. By this method it is possible to form bilayer membranes which are stable for several hours. The electrical

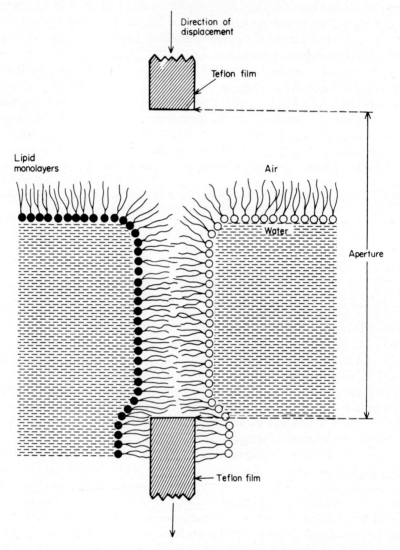

FIG. 18. The principle of phospholipid bilayer formation by the apposition of two previously formed monolayers at the air-water interface. The Teflon film of which the structural barrier between the two aqueous compartments is formed is a hydrophobic material, and so it gathers up the hydrophobic aspect of the lipid monolayers on to its surface as the level of the water in the troughs is raised. The lipid head groups remain in contact with the aqueous phase. The shading of the head groups of one monolayer indicates the possibility of forming asymmetric bilayers from monolayers of different composition. The figure is not drawn to scale. (From ref. 53.)

capacitance of such a membrane $(0\cdot9 \pm 0\cdot1 \text{ F.cm}^{-2})$ is about twice that of the typical solvent-containing bilayers formed by the brush technique. If we assume the usual value for the dielectric constant, then we find that the thickness of the resistive layer is 22 Å, i.e. 6–10 Å less than the thickness of biological membranes as determined by X-ray diffraction techniques (see Table 2). Actually, there is no intrinsic reason why the electrical thickness and the thickness as determined by other measurements should be precisely the same, but it is important to establish the reasons for the difference. As was pointed out earlier, calorimetric measurements show

TABLE 2.[53] A comparison of the electrical and dimensional properties of biological and phospholipid bilayer membranes prepared in the presence and absence of hydrocarbon solvents (references in brackets)

Measured property	Phospholipid bilayer containing hydrocarbon solvent (brush technique)	Phospholipid bilayer containing no hydrocarbon solvent (monolayer technique)	Cell membranes
DC resistance (ohm cm^2)	10^6–10^8 [12, 54]	10^6–10^8 [53]	$< 10^3$ [52]
DC resistance of gramicidin-treated membrane (see chapter 3)	10^3 [12, 54]	10^3 [53]	$< 10^3$ [55]
Membrane capacity (μF cm^{-2})	$0\cdot45 \pm 0\cdot05$ [6, 12, 54]	$0\cdot9 \pm 0\cdot1$ [53]	$0\cdot8$–$1\cdot2$ [6]
Thickness of hydrocarbon region (a) from capacitance measurements	42 Å	22 Å	16–24 Å
(b) from X-ray diffraction	—	29 Å	28–35 Å [57,58,59]

that a considerable amount of water is intimately associated with the lipid in a single lipid phase, and it has been suggested that it is the penetration of water into the superficial region of the hydrocarbon which reduces the effective electrical thickness of the phospholipid bilayer below that established by diffraction methods. These detect the reflections due to the heavier phosphorus atoms of the lipid head groups, but are unable to "see" the water.

Apart from representing a more perfect solution to the problem of modelling biological membranes, the solvent-free bilayers offer the possibility of greater versatility in certain respects. By forming the initial monolayers from different phospholipids, it is possible to form asymmetrical bilayers having different phospholipids on either half. Another possibility is that it may be easier to form bilayers from mixtures of proteins and lipids. One of the chief aims of current membrane biochemistry is to isolate and purify membrane proteins having enzyme activity, and then to reconstitute them in model membranes. Whilst a number of proteins have by now been isolated from membranes, only in a few instances has it been possible to effect this reconstitution, and the problem probably lies in the difficulty of penetrating a preformed bilayer with hydrated protein from the aqueous phase. A better approach may be to form the initial monolayer from a suitable mixture of phospholipid and purified protein before forming the bilayer. The protein could be added to either one or both monolayer surfaces, and in this way the orientation of the protein in the bilayer could be controlled. Inevitably, most studies with bilayers so far reported, derive from bilayers made with the much simpler brush technique, but there is an increasing awareness of the advantages of precision and versatility offered by the solvent-free membranes.

Permeation Mechanisms of Phospholipid Bilayer Membranes

The purpose of this section is to show that the mechanism of permeation of bilayers and biological membranes by water and small hydrophilic non-electrolytes might be similar.[60, 61]

The permeability of membranes to non-electrolytes can be measured by two independent approaches.

(a) Tracer Measurements at Equilibrium

In this method there is no net volume flow of material across the membrane. The membrane is formed to separate a solution of a given solute into two compartments, and the experiment is started by adding, to one compartment only, a trace quantity of the same solute bearing a radioactive label. The appearance of the radioactive material on the distal side of the membrane is measured by sampling at suitable intervals for counting, and the diffusion permeability coefficient $(P_d)_i$ of the radio-labelled solute is determined from the relationship

$$*\Phi_i = -(P_d)_i \, A(*c_{i_2} - *c_{i_1}),$$

where $*\Phi_i$ is the flux (moles/unit time) of the radioactive labelled solute, i

A is the area of the membrane

$*c_{i_1}$ and $*c_{i_2}$ are the concentrations of the solute on either side of the membrane.

If, instead of adding a labelled solute, a trace of tritiated water (^3H_2O) is added, then measurements of water permeability can be made.

(b) *Osmotic Measurements*

By this approach, a concentration gradient of a solute is established across the membrane and material then moves across the barrier until equilibrium is attained. Since the unilateral addition of a solute on one side of the membrane establishes an osmotic pressure gradient, water will tend to move in the opposite direction, and in the extreme case of an impermeant

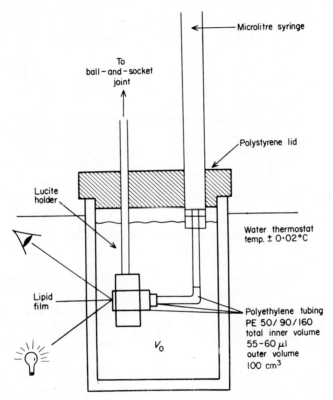

Fig. 19. Cell used for measuring osmotic flux across lipid bilayer membranes. Reproduced by permission of Cass *et al.* (1967). *J. Gen. Physiol.* **50**, 1765.

solute such as NaCl, the rate of flow of material will be limited only by the permeability characteristics of the membrane towards water. The greater the flux of the solute, the less water has to flow in order to re-establish osmotic equilibrium, and this variation forms the basis of solute permeability measurements across phospholipid bilayer membranes under non-equilibrium conditions. Typically, the membrane is formed at the end of a small polyethylene tube, and separates the solution contained in the tube from the solution contained in the external container. If a solute is now added to the external solution, there will be a movement of water across the membrane from the tube, tending to negate the change in osmotic pressure. This will result in a change in the hydrostatic pressure exerted on the membrane and distortion of its planar surface which can be detected optically. The adjustment of a micrometer-controlled microlitre syringe fitted to the free end of the polyethylene tube to restore planarity provides the means of measuring volume flow through the membrane. The method is sensitive to movements of as little as $0\cdot005$ μl.

If the osmotic gradient arises from an impermeant solute, then the volume flow will be due to water alone. The osmotic, or filtration permeability coefficient of water, P_{f_w}, is determined from the relationship

$$\Phi_w = -P_{f_w} A[\phi_i c_{i_2} - \phi_i c_{i_1}]$$

where Φ_w is the flux of water (moles/unit time)

A is the area of the membrane

ϕ_i is the osmotic coefficient of the impermeant solute

c_{i_1} and c_{i_2} are the concentrations of the impermeant solute on either side of the membrane,

thus $\Phi_w = -P_{f_w} A \phi \Delta c_i.$

If the gradient is due to a permeant solute, the resulting volume flux $\mathcal{J}_v$ of water will be reduced by the fraction σ_i where

$$(\mathcal{J}_v) \text{ impermeant} = \sigma_i \, (\mathcal{J}_v) \text{ permeant}$$
$$\text{solute} \qquad\qquad\qquad \text{solute}$$

σ_i, referred to as the reflection coefficient of the solute, is a measure of the permeability of the membrane towards that solute.

With the approaches outlined above, it is possible to determine two permeability coefficients for water, P_{d_w} and P_{f_w} at equilibrium and under the influence of an osmotic gradient, and two permeability coefficients for solutes, P_{d_i} and σ_i.

These parameters, P_{d_w} and P_{d_i} on the one hand, and P_{f_w} and σ_i on the other, are all coefficients of membrane permeability, but the physical basis underlying their significance is quite different.[60, 61, 62, 63] The difference can be most clearly invisaged in the limiting conditions of zero concentration gradient in which case diffusion is maintained in the absence of any volume flow.

The relationships between these parameters can be used to determine the mechanism of permeation through the membrane, thus, the condition of $P_{d_w} < P_{f_w}$ is indicative of the existence of pores, which have to be filled with water in order for water to flow (the water molecules are "aware of each other" and move in a cooperative manner). Similarly, the condition of $\sigma_i < 1$ is indicative of interaction of the solute, i, with water as it traverses the membrane through an aqueous channel. Conversely, when $P_{d_w} = P_{f_w}$ (or $\sigma_i = 1$) it is generally concluded that the permeation process involves the transfer of single molecules by a mechanism of solvation and diffusion through the lipid barrier.

WATER PERMEABILITY OF PHOSPHOLIPID BILAYER MEMBRANES

Early attempts to measure water fluxes through phospholipid bilayers produced results suggestive of a common mechanism for water permeation with biological membranes. Firstly, the actual magnitude of the measured fluxes (see Table 3, first two rows) appeared similar,[64, 65] and secondly, the measurements suggested that the diffusion coefficient is substantially less than the filtration coeficient. A similar relationship between P_{d_w} and P_{f_w} is readily discerned in water flux measurements across artificial barriers formed from cellulose[66] or collodion,[67] in which the transport mechanism is clearly by way of water-filled pores, and the finding of $P_{d_w} < P_{f_w}$ was widely accepted among physiologists as being diagnostic of pore mediated transport. From the outset it was generally recognised that this would be a doubtful interpretation in the case of phospholipid bilayer membranes.

Whilst accepting the possibility of unity of mechanism, it has been pointed out that there is a fundamental flaw in the argument which leads to the conclusion that water permeation occurs by way of water filled channels.[70] The problem arises from the existence of unstirred layers in the solvent, which can extend up to distances of a few hundreds of micrometres from the surface of a membrane into the bulk phase. The analysis of volume flow measurements under a pressure gradient leading to the estimate of P_f are unlikely to be complicated by artefacts of this sort, but it must be apparent that diffusional flux measurements of a tracer will be subject to drastic underestimation due to the presence of unstirred layers which must lead to depletion of the tracer at one surface and its accumulation at the other.

The data recorded in line 3 of Table 3 arise from experiments in which strenuous efforts were taken in order to maximise stirring on both sides of the bilayer in the tracer diffusion experiment.[61] Under these conditions,

TABLE 3. Comparison of P_{d_w} and P_{f_w} for phospholipid bilayers and some biological membranes.

	P_{d_w} cm/sec 10^{-3}	P_{f_w} cm/sec 10^{-3}	Ratio	Ref.
phospholipid bilayers	0·23	0·88–1·99	4–8	64
	0·44	1·7–10·4	4–24	65
	1·06	1·14	1·1	61
red blood cells	5·3		2·5	68
invertebrate peripheral nerve fibres	0·4		20	69

$P_d \sim P_f$. Similarly, for hydrophilic solutes (acetamide, urea) $\sigma_i = 1$ indicating no interaction with water during passage through the membrane. These results for the permeation of phospholipid bilayers by water and small hydrophilic solutes are diagnostic of a mechanism involving solvation and diffusion of individual molecules in the lipid phase.[60]

It is relevant at this point to ask whether water transport in biological cell membranes might also occur by a solubility-diffusion mechanism, and not, as has often been suggested, by way of aqueous pores. In the case of biological membranes, the evidence for aqueous pores rests firmly on the finding that $P_{d_w} < P_{f_w}$ (see Table 3). However, bearing in mind the special precautions required to eliminate unstirred layer artefacts on bilayers (with both surfaces accessible), one can understand that measurements with cells are likely to provide severe underestimates of P_{d_w} and the possibility of drawing false conclusions about their mechanism.[70] A better approach to the question of water transport in biological membranes is to consider other parameters of the solubility-diffusion mechanism which can be applied to both model and biological membranes.

One such approach has been to show that the Arrhenius activation energy for the water transport process is similar for both systems (this is determined from measurements of permeability at different temperatures and recording the slope of the graph of log Φ vs $1/T°$ abs). In the case of phosphatidyl choline bilayers the value of the experimental activation energy E_a (about 12 Kcal mole^{-1}), is close to the value that may be predicted on theoretical grounds for a solubility-diffusion mechanism in which E_a is composed of two parts: the activation energy of diffusion (3–4·5 Kcal mole^{-1}) and the enthalpy of the partition coefficient of water between the bulk aqueous phase and the membrane (8–9 Kcal mole^{-1}).[71, 72]

The antidiuretic hormone (ADH or vasopressin) stimulates water and Na$^+$ movements in kidney and the skin of amphibia. ADH reacts with phospholipid bilayers to reduce the energy of activation (E_a) of water movement to about 4 Kcal mole^{-1} and to increase flow, especially at lower temperatures (20°; at 37° there is little effect).[73, 74] There is no effect of ADH on Na$^+$ transport in bilayers, but the resemblance in magnitude and effect on water transport on bilayers and some biological membranes is close. As for biological membranes, the effects of ADH is highly specific: the closely related structure, oxytocin is inactive in mediating water movement.

Is the idea of aqueous solution in a hydrocarbon phase a realistic one?[75] The solubility of water in hexadecane at 35° is 64 mg l^{-1}, i.e. 3·5 mM,[76] and the diffusion coefficient is 5.10^{-5} cm^2 sec^{-1}. If the thickness of the hydrocarbon region of the black lipid membrane is taken as 50 Å, we get $P_f = 6·4.10^{-3}$ cm sec^{-1}; which is very close to the measured value of $P_f = 4·2.10^{-3}$ cm sec^{-1} for phosphatidyl choline bilayers. The permeability of bilayers to water is reduced when cholesterol is present in the phospholipid mixture used to prepare them. This reduction in permeability probably arises from the decreased fluidity of the hydrocarbon phase described earlier in this chapter.[72, 24]

All these observations argue for a solubility diffusion mechanism of water transport across phospholipid membranes and raise the strong possibility of a similar mechanism operating in biological membranes.

ELECTROLYTE PERMEABILITY OF PHOSPHOLIPID BILAYER MEMBRANES

The permeability of phospholipid bilayer membranes to cations is about 10^9–10^{10} times less than the permeability to water,[3] and it is very likely that the very low permeability towards ions arises from the low solubility of ions in the hydrocarbon phase. The low solubility can be predicted on thermodynamic grounds (or energy of charging): the self energy[77] of a spherical univalent ion in a given medium is

$$E = e^2/2\epsilon r,$$

where e = electronic charge
ϵ = dielectric constant.
r = radius

For fully hydrated Na^+ of corrected Stokes radius = 3.3 Å in water, E = 0.6 Kcal mole^{-1} (~ 1 RT at $25°$). In a hydrocarbon medium, where $\epsilon = 2$, $E = 25$ Kcal mole^{-1} (41 RT at $25°$). On this basis, the free energy increase ΔG on transporting one mole of Na^+ from water to the interior of a phospholipid bilayer will be about 40 RT. Now, one can calculate the concentration of Na^+ in the interior of the bilayer from the Boltzmann distribution if the Na^+ concentration in the aqueous phase is known:

$$C_{org} = C_{aq} \exp{-(\Delta G/RT)},$$

so if the concentration of Na^+ in the aqueous phase is 0.1 M, the concentration of Na^+ in the interior of the bilayer will be about 10^{-19} M!

The reasons for the wide disparity in ion transport properties between lipid bilayers and cell membranes are quite clear. In complete cell membranes there are inserted into the hydrocarbon phase molecular pathways which regulate ion flow. In essence, their task is to increase the effective solubility of ions in the membrane. This can be done either by increasing the dielectric constant in the region of the membrane through which the ion passes (i.e. by forming a water-filled channel or pore), or by extensive charge delocalisation to increase r, the radius of the charged species (by complexation with specialised ionophoric carrier molecules). These pathways lie in parallel with the highly resistive hydrocarbon barrier and, as a result of this, any contribution the hydrocarbon makes to ionic conductance is negligible. In Chapter 3, we shall explore some of the ways in which these highly resistive barriers have been modified to make them resemble cell membranes more closely.

REFERENCES

1. *Bangham, A. D. (1968). Membrane models with phospholipids. *In* "Progress in Biophysics and Molecular Biology", J. A. V. Butler and D. Noble (eds). Pergamon Press, Oxford.
2. *Bangham, A. D. (1972). Lipid bilayers and biomembranes. *Ann. Rev. Biochem.* **41**, 753.
3. *Mueller, P. and Rudin, D. O. (1969). Translocators in bimolecular lipid membranes: their role in dissipative and conservative bioenergy trans- ductions. *In* "Current Topics in Bioenergetics, Vol. 3", D. R. Sanadi (ed). Academic Press, New York and London.
4. Goldacre, R. J. (1958). Surface films, their collapse on compression, the shapes and sizes of cells and the origin of life. *In* "Surface Phenomena in Chemistry and Biology", J. F. Danielli, K. G. A. Pankhurst and A. C. Riddiford (eds). Pergamon Press, Oxford.
5. Overton, E. (1899). The probable origin and physiological significance of cellular osmotic properties (a translation by R. B. Park of the original "Ueber die allgemeinen osmotischen Eigenschaften der Zelle, ihr vermutlichen Ursachen und ihre Bedeutung für die Physiologie"). *In* "Papers on Biological Membrane Structure", D. Branton and R. B. Park (eds). Little, Brown and Co., Boston (1968).
6. Fettiplace, R., Andrews, D. M. and Haydon, D. A. (1971). The thickness, composition and structure of some lipid bilayers and natural membranes. *J. Membrane Biol.* **5**, 277.
7. Chapman, D. (1969). "Introduction to Lipids." McGraw-Hill Publishing Company, London.
8. Gorter, E. and Grendel, F. (1925). On bimolecular layers of lipoids on the chromocytes of the blood. *J. Exp. Med.* **41**, 439.
9. Davies, J. T. and Rideal, E. K. (1961). "Interfacial Phenomena." Academic Press, New York and London.
10. van Deenen, L. L. M., Houtsmuller, U. M. T., de Haas, G. H. and Mulder, E. (1962). Monomolecular layers of synthetic phosphatides. *J. Pharm. Pharmacol.* **14**, 429.
11. Demel, R. A., Bruckdorpher, K. R. and van Deenen, L. L. M. (1972). Structural requirements of sterols for the interaction with lecithin at the air-water interface. *Biochem. Biophys. Acta*, **255**, 311.
12. Bangham, A. D., Standish, M. M. and Watkins, J. C. (1965). Diffusion of univalent ions across the lamellae of swollen phospholipids. *J. Mol. Biol.* **13**, 238.
13. Luzzati, V. and Husson, F. (1962). The structure of the liquid-crystalline phases of lipid water systems. *J. Cell. Biol.* **12**, 207.
14. Stoeckenius, W. (1962). Some electron microscopical observations on liquid crystalline phases in lipid-water systems. *J. Cell. Biol.* **12**, 221.
15. Luzzati, V. (1968). X-ray diffraction studies of lipid-water systems. *In* "Biological Membranes, Physical Fact and Function", D. Chapman (ed). Academic Press, London and New York.
16. *Bangham, A. D., Hill, M. W. and Miller, N. G. A. (1974). Preparation and

* References marked with an asterisk (*) are mainly review articles especially recommended for further reading.

use of liposomes as models of biological membranes. *Methods in Membrane Biology*, **1**, 1.

17. Fernandez-Moran, H. (1962). New approaches in the study of biological ultrastructure by high resolution electronmicroscopy. *In* "The Interpretation of Ultrastructure", R. J. C. Harris (ed). Academic Press, London and New York.

18. Bangham, A. D. and Horne, R. W. (1964). Negative staining of phospholipids and their structural modification by surface active agents as observed in the electron microscope. *J. Mol. Biol.* **8**, 660.

19. Papahadjopoulos, D. and Miller, N. (1967). Phospholipid model membranes: structural characteristics of hydrated liquid crystals. *Biochim. Biophys. Acta*, **135**, 624.

20. Bangham, A. D., de Gier, J. and Greville, G. D. (1967). Osmotic properties and water permeability of phospholipid liquid crystals. *Chem. and Phys. of Lipids*, **1**, 46.

21. Bangham, A. D., Pethica, B. A. and Seaman, G. V. F. (1958). The charged groups at the interface of some blood cells. *Biochem. J.* **69**, 12.

22. Bangham, A. D. and Papahadjopoulos, D. (1966). Interaction of phosphatidylserine monolayers with metal ions. *Biochim. Biophys. Acta*, **126**, 181.

23. Johnson, S. M., Bangham, A. D., Hill, M. W. and Korn, E. D. (1971). Single bilayer liposomes. *Biochim. Biophys. Acta*, **233**, 820.

24. *Chapman, D. (1968). Recent physical studies of phospholipids and natural membranes. *In* "Biological Membranes, Physical Fact and Function", D. Chapman (ed). Academic Press, London and New York.

25. *Chapman, D. (1973). Some recent studies of lipids, lipid-cholesterol and membrane systems. *In* "Biological Membranes", Vol. 2, D. Chapman and D. F. H. Wallach (eds). Academic Press, London and New York.

26. Chapman, D., Williams, R. M. and Ladbrooke, B. D. (1967). Physical studies of phospholipids: thermotropic and lyotropic mesomorphism of some 1.2-diacylphosphatidyl cholines (lecithins). *Chem. and Phys. of Lipids*, **1**, 445.

27. Williams, R. M. and Chapman, D. (1970). Phospholipids, liquid crystals and cell membranes. *Prog. Chem. Fats and other Lipids*, **XI**, 1.

28. Chapman, D., Byrne, P. and Shipley, G. G. (1966). The physical properties of some 2.3-diacyl-DL-phosphatidylethanolamines. *Proc. Roy. Soc. A*, **290**, 115.

29. Chapman, D. and Salsbury, N. J. (1966). Physical studies of phospholipids: proton magnetic resonance studies of molecular motion in some 2.3-diacyl-DL-phosphatidylethanolamines. *Trans. Faraday Soc.* **62**, 2607.

30. Chapman, D., Keough, K. M. and Urbina, J. (1974). Biomembrane phase transitions: studies of lipid-water systems using differential scanning calorimetry. *J. Biol. Chem.* **249**, 2512.

31. Ladbrooke, B. D., Williams, R. M. and Chapman, D. (1968). Studies on lecithin-cholesterol-water interactions by differential scanning calorimetry and X-ray diffraction. *Biochim. Biophys. Acta*, **150**, 333.

32. Chapman, D. and Morrison, A. (1966). Physical studies of phospholipids: high resolution nuclear magnetic resonance spectra of phospholipids and related substances. *J. Biol. Chem.* **241**, 5044.

33. Chapman, D. and Penkett, S. A. (1966). Nuclear magnetic resonance spectroscopic studies of the interaction of phospholipids with cholesterol. *Nature*, **211**, 1304.

34. Ladbrooke, B. D. and Chapman, D. (1969). Thermal analysis of lipids, proteins and biological membranes: a review and summary of some recent studies. *Chem. and Phys. of Lipids*, **3**, 304.

35. Steim, J. M. (1970). Thermal phase transitions in biomembranes. *In* "Liquid Crystals and Ordered Fluids", J. F. Johnson and R. S. Porter (eds). Plenum Press, New York.

36. Chapman, D. and Urbina, J. (1971). Phase transitions and bilayer structure of *Mycoplasma laidlawii*. *FEBS Lett.* **12**, 169.

37. Overath, P., Schairer, H. U. and Stoffel, W. (1970). Correlation of *in vivo* and *in vitro* phase transitions of membrane lipids in *Escherichia coli*. *Proc. Nat. Acad. Sci.* **67**, 606.

38. McElhaney, R. N. (1974). The effect of membrane-lipid phase transitions on membrane structure and on the growth of *Acholesplasma laidlawii*. *J. Supramolecular Structure*, **2**, 617.

39. Kimelberg, H. K. and Papahadjopoulos, D. (1972). Phospholipid requirements for $(Na^+ + K^+)$-ATPase activity: head group specificity and fatty acid fluidity. *Biochim. Biophys. Acta*, **282**, 277.

40. Tsukagoshi, N. and Fox, C. F. (1973). Abortive assembly of the lactose transport system in *Escherichia coli*. *Biochemistry*, **12**, 2816.

41. de Kruijff, B., van Dijck, P. W. M., Goldbach, R. W., Demel, R. A. and van Deenen, L. L. M. (1973). Influence of fatty acid and sterol composition on the lipid phase transition and activity of membrane bound enzymes in *Acholeplasma laidlawii*. *Biochim. Biophys. Acta*, **330**, 269.

42. Machtiger, N. A. and Fox, C. F. (1973). The biochemistry of bacterial membranes. *Ann. Rev. Biochem.* **42**, 575.

43. Sinensky, M. (1974). Homeoviscous adaptation: a homeostatic process that regulates the viscosity of membrane lipids in *Escherichia coli*. *Proc. Nat. Acad. Sci.* **71**, 522.

44. Huang, L., Lorch, S. K., Smith, G. G. and Haug, A. (1974). Control of membrane lipid fluidity in *Acholeplasma laidlawii*. *FEBS Lett.* **43**, 1.

45. Papahadjopoulos, D. and Watkins, J. C. (1967). Phospholipid membranes: permeability properties of hydrated liquid crystals. *Biochim. Biophys. Acta*, **135**, 639.

46. de Gier, J., Mandersloot, J. G. and van Deenen, L. L. M. (1969). The role of cholesterol in lipid membranes. *Biochim. Biophys. Acta*, **173**, 143.

47. Demel, R. A., Bruckdorfer, K. R. and van Deenen, L. L. M. (1972). The effect of sterol structure on the permeability of liposomes to glucose, glycerol and Rb^+. *Biochim. Biophys. Acta*, **255**, 321.

48. Newton, I. (1704). "Optiks". Facsimile edition by Dover Publications Inc., New York, 1952.

49. Mueller, P., Rudin, D. O., Tien, H. T. and Wescott, W. C. (1962). Reconstitution of excitable cell membrane structure *in vitro*. *Circulation*, **26**, 1167.

50. Mueller, P. and Rudin, D. O. (1969). Bimolecular lipid membranes: techniques of formation, study of electrical properties, and induction of ionic gating phenomena. *In* "Laboratory Techniques in Membrane Biophysics: an Introductory Course", H. Passow and R. Stampfli (eds). Springer-Verlag, Berlin.

51. Mueller, P., Rudin, D. O., Tien, H. T. and Wescott, W. C. (1964). Formation and properties of bimolecular lipid membranes. *In* "Recent Progress in Surface Science", Vol. 1, J. F. Danielli, K. G. A. Pankhurst and A. C. Riddiford (eds). Academic Press, London and New York.

52. Cole, K. S. (1968). "Membranes, Ions and Impulses." University of California Press, Berkeley, California.

53. Montal, M. and Mueller, P. (1972). Formation of bimolecular membranes from lipid monolayers and a study of their electrical properties. *Proc. Nat. Acad. Sci.* **69**, 3561.

54. *Thomson, T. E. and Henn, F. A. (1970) Experimental phospholipid model membranes. *In* "Membranes of Mitochondria and Chloroplasts" (ACS Monograph No 165), E. Racker (ed). van Nostrand, New York.

55. Chance, B. and Montal, M. (1971). Ion translocation in energy-conservation membrane systems. *In* "Current Topics in Membranes and Transport", Vol. 2, F. Bronner and A. Kleinzeller (eds). Academic Press, New York.

56. Levine, Y. K. and Wilkins, M. H. F. (1971). Structure of oriented lipid bilayers. *Nature New Biol.* **230**, 69.

57. Engelman, D. (1969). Surface area per lipid molecule in the intact membrane of the human red cell. *Nature*, **223**, 1279.

58. Engelman, D. (1970). X-ray diffraction studies of phase transitions in the membrane of *Mycoplasma laidlawii*. *J. Mol. Biol.* **47**, 115.

59. Wilkins, M. H. F., Blaurock, A. E. and Engelman, D. M. (1971). Bilayer structure in membranes. *Nature New Biol.* **230**, 72.

60. Finkelstein, A. and Cass, A. (1968). Permeability and electrical properties of thin lipid films. *J. Gen. Physiol.* **52**, suppl. 145.

61. Cass, A. and Finkelstein, A. (1967). Water permeability of thin lipid membranes. *J. Gen. Physiol.* **50**, 1765.

62. *Dick, D. A. T. (1965). "Cell Water". Butterworth, London.

63. *Stein, W. D. (1967). "The Movement of Molecules across Cell Membranes." Academic Press, New York and London.

64. Hanai, T., Haydon, D. A. and Taylor, J. (1965). Some further experiments on bimolecular lipid membranes. *J. Gen. Physiol.* **48** suppl., 59.

65. Huang, C. and Thompson, T. E. (1966). Properties of lipid bilayer membranes separating two aqueous phases: water permeability. *J. Mol. Biol.* **15**, 539.

66. Renkin, E. M. (1954). Filtration, diffusion and molecular sieving through porous cellulose membranes. *J. Gen. Physiol.* **38**, 225.

67. Robbins, E. and Mauro, A. (1960). Experimental study of the independence of diffusion and hydrodynamic permeability coefficients in collodion membranes. *J. Gen. Physiol.* **43**, 523.

68. Paganelli, C. V. and Solomon, A. K. (1957). The rate of exchange of tritiated water across the human red cell membrane. *J. Gen. Physiol.* **41**, 259.

69. Nevis, A. H. (1957). Water transport in invertebrate peripheral nerve fibres. *J. Gen. Physiol.* **41**, 927.

70. Dainty, J. (1963). Water relations of plant cells. *Adv. Botan. Res.* **1**, 279.

71. Price, H. D. and Thompson, T. E. (1969). Properties of liquid bilayer membranes separating two aqueous phases: temperature dependence of water permeability. *J. Mol. Biol.* **41**, 443.

72. Graziani, Y. and Livne, A. (1972). Water permeability of bilayer lipid membranes: sterol-lipid interaction. *J. Membrane Biol.* **7**, 275.

73. Graziani, Y. and Livne, A. (1971). Vasopressin and water permeability of artificial lipid membranes. *Biochem. Biophys. Res. Com.* **45**, 321.

74. Graziani, Y. and Livne, A. (1973). Bilayer lipid membrane as a model for vasopressin; prostaglandin and Ca^{2+} effects on water permeability. *Biochim. Biophys. Acta*, **291**, 612.

75. Schatzberg, P. (1965). Diffusion of water through hydrocarbon liquids. *J. Polymer Sci.* **10**, pt C, 87.

76. Schatzberg, P. (1963). Solubilities of water in several normal alkanes from C_7 to C_{16}. *J. Phys. Chem.* **67**, 776.

77. Parsegian, A., (1969). Energy of an ion crossing a low dielectric membrane: solutions to four relevant electrostatic problems. *Nature*, **221**, 844.

2 | The Structure of Biological Membranes

EARLY EVIDENCE FOR MOLECULAR ORGANISATION IN BIOLOGICAL MEMBRANES

It was the spreading properties of fatty extracts of red blood cells at the air-water interface that first led to the proposal that lipids might be organised as a molecular bilayer within membranes.[1] We have seen that the bilayer represents a thermodynamically stable organisation for phospholipid molecules, and that bilayer structures, devoid of proteins, possess a number of the linear or passive properties which may be said to typify membranes. Nonetheless, a great deal more evidence is needed before we can legitimately accept the idea of the phospholipid bilayer even as the central structural feature of the red blood cell membrane. If we wish to extend the principal of the lipid bilayer to other membrane systems, then we should bear in mind the many roles in which membranes play a part (ranging from the rather inert barriers of myelin to the catalytic centres of such highly organised activities as photosynthesis and oxidative phosphorylation). We should ask if it is appropriate that such disparate activities should be controlled by a single basic structure.

We might ask if it is possible for components of membranes other than lipids to assemble in a suitable manner to provide membrane-like structures. Indeed, the outer coat of the bacteriophage head may be said to be a pure protein membrane. The protein membrane as revealed by electron microscopy is about 60 Å thick, but does not afford the characteristic trilaminar appearance of a biological cell membrane. Model protein vesicles have been prepared, generally from mixtures of cationic and anionic synthetic polypeptides (e.g. polyaspartic acid and polylysine). These do not behave as osmometers and have very low electrical resistance (of the order of 100 Ω cm²).[2] Certainly protein membranes would not appear to have the barrier properties which are the first and most obvious feature of all living membranes.

Thus it is worth returning to the idea of the lipid bilayer, to examine the evidence for its existence in biological membrane structure.

THE AREA PROBLEM

The mammalian red blood cell remains the only example where the identity of the areas of the cell membrane surface and lipid extract mono-layer preparations have been systematically compared (see Chapter 1, p. 8). Whilst the original bilayer proposal of Gorter and Grendel[1] stimulated a wealth of investigation, it would be true to say that the experiments upon which their ideas were based are open to criticism from three points of view.

(1) Although it is probably true that all the lipid of the red cell is contained within the membrane, it would be essential in experiments of this sort to achieve something approaching 100% extraction: the method used by Gorter and Grendel was extraction with acetone, which gives significantly less than the total.

(2) The surface area of the extracted lipids as a monolayer was compared with the surface area of red cells in dried film preparations (99 μ^2 instead of 145 μ^2 in wet films). Thus the original experiments were based on an underestimate of both the amount of lipid and the surface area to be covered, and fortuitously provided the 2:1 ratio between the extract monolayer and red cell surface areas.

(3) The outcome of experiments of this sort depends upon the more serious assumption that the area occupied by the lipids in the membrane is the same as the area occupied by lipids on a spread monolayer in a Langmuir trough at the point at which the first increment in surface pressure due to the presence of the lipids becomes detectable.

A recent re-examination of this experiment,[3] in which care was taken to achieve total extraction of red cell lipid, and in which areas could be compared with wet film areas of red cells, has shown that the 2:1 ratio of spread monolayer to cell surface area is tenable so long as low film pressures (5–15 dyne cm^{-1}) can be assumed. If the film pressure is increased, collapse of the film occurs well before a 1:1 ratio is attained. This certainly disposes of any possibility of a monomolecular arrangement of lipids in the membrane, but it can hardly be said to be a proof of bilayer organisa-tion. It must be admitted that the great importance of the original experi-ment was that it led Gorter and Grendel on to the germinal principle which has dominated all thinking and controversy about membrane structure ever since.

SURFACE TENSION

Historically, a necessary role for proteins or surface active material on membrane surfaces was provided on the basis of the great disparity which exists between the tension at membrane-water interfaces (about 0.6 dyne cm^{-1}) and oil-water interfaces (> 10 dyne cm^{-1}). The finding that the surface tension of oil droplets coated with protein (actually, the oil droplets which cause buoyancy in mackerel eggs) is similar to that of some membranes led to the first proposals for lipid-protein organisation in membranes[4] and the first membrane model of Danielli and Davson[5, 6, 7] (Fig. 1) is simply an extension of the "oil drop-protoplasm" interface which had been considered with regard to interfacial tension. The chief barrier is

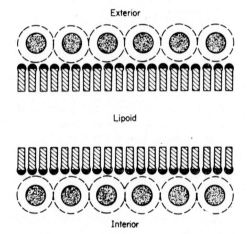

FIG. 1. The original "model" of membrane organisation: the proposal of Davson and Danielli consisted of a thin sheet of lipoid of undefined thickness, having protein adsorbed on both surfaces. (From ref. 7.)

seen to be a continuous lipid film of undefined thickness, but having clearly oriented molecular layers at the surface. To these surfaces there is adsorbed a "molecular mosaic of dense impenetrable areas interspersed by heavily hydrated layers of molecular dimensions, the whole being a relatively permanent structure". Penetration of the film was considered to occur both by solution in the film, and through pores, which could exist as permanent entities in solid films. Selectivity amongst charged molecules would be determined by the isoelectric properties of the protein film.

These remarks, made forty years ago, were the first to direct attention to the possible role and disposition of protein in membranes, and it is only in very recent times that alternative arrangements for membranes have been seriously considered. The pertinance of this discussion is highlighted by experience with the black lipid bilayer membranes, for when these were first described, they were found to have a surface tension on the high side of the range for membranes of biological origin. It then looked as if the original reasons for invoking a role for protein as a component of membrane structure had been erroneous. However, recent experience (Chapter 1, p. 37) has shown that solvent-free lipid bilayers have a surface tension which is close to zero, and tend to expand spontaneously without any applied force. There is then something "special" about the biological membrane, and one is bound to respect the clear-sighted thinking which underlay the creation of the Davson and Danielli model: an idea which has in many respects stood the test of time and criticism.

ELECTRON MICROSCOPY

Three Layered Images and the Unit Membrane

If the idea of a continuous lipid bilayer having protein adsorbed on the surface is tenable for the mammalian red cell membrane, is it logical to impose this idea as the central structural feature of all other membranes? The general concept in its subsequent development came to be known as the "unit membrane" hypothesis.[8, 9, 10, 11, 12]

One of the most compelling pieces of evidence which supported the idea of a common general structure for membranes is the existence of a three-layered image, almost invariably observed in electron microscope pictures of membranes, which have been fixed and stained, and sectioned normal to the plane of the membrane. Figure 2 shows clearly the trilaminar appearance of the sectioned red cell membrane. Intracellular organelle membranes, too, yield the trilaminar image.

The familiar trilaminar image produced by electron microscopy of biological membranes brings us face to face with a number of problems of interpretation. These are made all the more problematic by virtue of the fact that a trilaminar image is just what one might seek when looking for electron microscopic evidence of molecular bilayer structure; a preconceived idea is readily imposed upon what is observed, and it is all too easy to be blinded by the resulting mirage of clarity.

Fixation for Electron Microscopy—and some of its Problems

The preparation of a tissue specimen for electron microscopy typically involves a number of processes including fixation, dehydration, staining, embedding and sectioning. We should consider how some of these procedures might affect the molecular arrangements within the membrane itself.[13, 14]

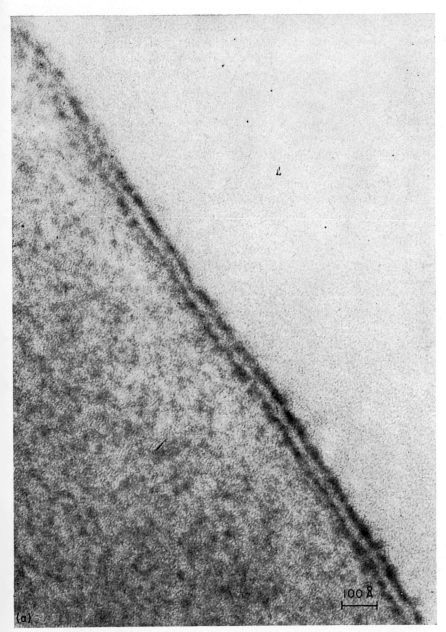

FIG. 2. Electron micrograph of a portion of a human red blood cell, showing the trilaminar appearance associated with unit membrane structure at its surface. The cell was fixed with KMnO₄. Reproduced by permission of Robertson (1964) from "Cellular Membranes", ed. Locke. Academic Press.

C

It is usual nowadays to fix tissues initially in a solution of glutaralde-hyde, though formerly other fixatives, such as potassium permanganate, osmic acid and formalin-dichromate, were used. Glutaraldehyde reacts with primary amino groups with great rapidity, and it is understood that it forms crosslinks between the free amino groups of proteins. Quite ob-viously, glutaraldehyde is a potent "denaturant" and destroys the native arrangement of all proteins with which it reacts. It is generally held that glutaraldehyde has little effect on the final appearance of the electron microscope image of membranes, but of course, any uncertainty adds to the problems of interpreting the final image. Postfixation is done in solu-tions of osmium tetroxide OsO_4. As well as having effects on the molecular organisation of the membranes, there is the possibility that the Os atom (at. wt. 195) could contribute to the electron density of the final sectioned preparation; it is therefore very important to judge its final location within the structure. Fixed preparations are then dehydrated by successive transfer between solutions of increasing ethanol content, but the ethanol has to be exchanged for a non-polar solvent (commonly propylene oxide) in order to make the dehydrated specimen miscible with the epoxy resin (Araldite) which is used for embedding. The epoxy monomer resin is allowed to penetrate the specimen for a number of days, and it is then polymerised by heating at $60°$. Thin sections (down to about 500 Å) may be cut from the hardened block by the use of a diamond knife in an ultramicrotome and these can now be stained by immersion in solutions of heavy metals, such as uranyl acetate and lead citrate. Once again it is of the utmost importance to know what will be the final destination of these heavy metal stains. If it is the contribution of the lipid (i.e. the bilayer) to the final appearance of the electron microscopic image that we wish to examine, then we must remember that fixation with glutaraldehyde cannot rescue the lipid from the subsequent effects of dehydration in alcohol. This is why thin section electron microscopy has not been widely applied in the study of the phospholipid model membranes of Chapter 1. It is probable that OsO_4 treatment contributes both to the fixation of proteins and lipids. In model reactions with long chain unsaturated fatty acids (Fig. 3), esterification and crosslinking between neighbouring molecules oc-curs across the double bonds.[14] Another product of the reaction is the dioxide, OsO_2, but this does not bind to the lipids; its location in a tissue preparation is unknown, though it is possible that it could accumulate at lipid water interfaces, and contribute in an arbitrary way to the electron density of the three layered image of membranes.

There are other reasons for believing that the evidence of the electron microscope is insufficient to generate arguments in favour of any particular molecular arrangements in membranes. In particular, removal of the lipid component (which we have considered so far as the central unifying feature of membrane structure) from membrane preparations prior to fixation with glutaraldehyde or OsO_4 has remarkably little effect on the appearance of the final image. Thus, mitochondrial inner membranes which have been defatted by treatment with 95% acetone before fixation, still produce the

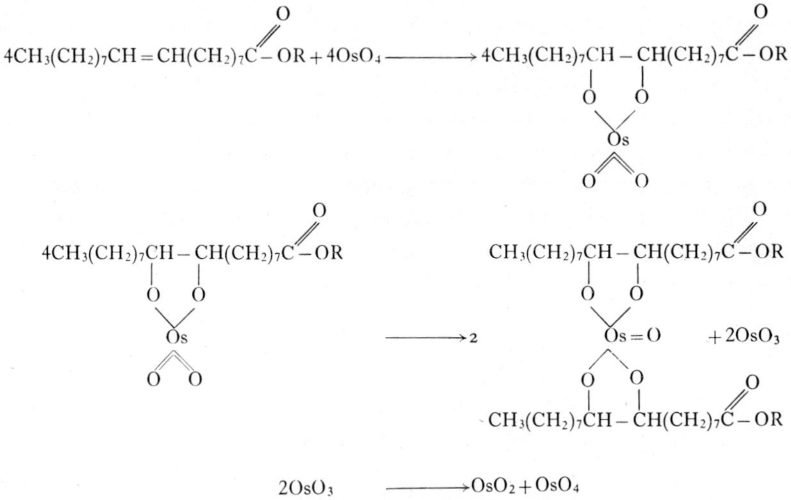

$$4CH_3(CH_2)_7CH = CH(CH_2)_7\overset{\displaystyle O}{\overset{\|}{C}} - OR + 4OsO_4 \longrightarrow 4CH_3(CH_2)_7CH - CH(CH_2)_7\overset{\displaystyle O}{\overset{\|}{C}} - OR$$

FIG. 3. The possible course of reaction between osmium tetroxide and unsaturated fatty acids. As well as "fixing" the unsaturated lipid, the reaction yields OsO_2, which may be deposited at the lipid-water interface. (From ref. 14.)

typical trilaminar image of the "bilayer".[15] This suggests that the material which constitutes a large fraction of the actual mass of the membrane, and which in large part provides the necessary environment for its reactivity, and all of its barrier properties, and which is likely to react with some of the fixatives and stains, is detected to quite an unknown extent in the normal processes of thin section electron microscopy.

X-RAY DIFFRACTION

In order to assess the damage suffered by the processes of fixing and staining for electron microscopy, an independent means of investigating the structural details of membranes is needed. This has been amply provided in the case of the repetitive membranes of myelin and retinal rod outer segments, by the techniques of X-ray diffraction—which yield much other important information besides.

The application of X-rays for structural determination depends on the original realisation that crystalline structures can behave as three-dimensional diffraction gratings, if the radiation is of sufficiently short wavelength. Every atom can scatter X-rays to an extent dependent on the number of orbital electrons, and every plane of atoms in the crystal behaves to X-rays (wavelength, $\lambda \sim 10^{-10}$ m) just as a mechanically ruled line in a diffraction grating to visible light (wavelength of green light $\sim$ $5 \cdot 5.10^{-7}$ m). Thus X-ray diffraction patterns provide information both about the spacial arrangement of atoms within a crystal, and about the electron density and hence the mass of the scattering atoms.

In Fig. 4 (*a*) the reflection of the ray AO, along the line OZ can be reinforced by or reduced by other parallel reflections depending on whether these emerge in or out of phase with each other. When the emergent rays, reflected from the different reflecting planes emerge exactly in phase, there is maximal reinforcement producing intense reflected radiation; when the emergent rays are exactly out of phase there is total cancellation. The diffraction picture (which can be photographed or recorded instrumentally)

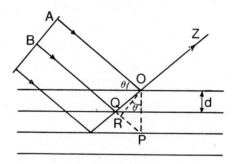

FIG. 4 (*a*). X-ray reflection from equidistant planes.

thus consists of an assemblage of spots of varying intensity. The condition for reinforcement is determined geometrically. A line OP is drawn perpendicular to the reflecting planes, and another line OR is drawn from O perpendicular to the incoming parallel rays. The angle ROP is equal to θ, the angle of incidence of the incoming rays. The pathlength difference for two rays AOZ and BQZ reflected from two adjacent atomic planes separated by the distance, d, is OQ $-$ QR, and because OQ $=$ QP, the pathlength difference is QP $-$ QR $= 2d \sin \theta$. To maximise reflection (i.e. for the two emergent rays to be in phase with each other) the pathlength difference must be an integral number n, of wavelengths, so $n\lambda = 2d \sin \theta$. This is the Bragg equation.

For a particular structural arrangement and source of X-rays, d and λ are fixed. The diffraction pattern thus consists of a series of intense reflections separated by areas of cancellation as the angle of incidence is varied. The reflections correspond to the integral values of $n = 1, 2, 3, 4 \ldots$ at which maximal reinforcement occurs. The larger the spacings within the structure, the smaller will be the distance between the spots of the diffraction pattern; thus, for membranes and phospholipid aggregates, two quite distinct patterns are formed. The low-angle pattern, close to the axis of the incoming collimated X-ray beam provides information about the spacings of the lamellar organisation (20–500 Å); the wide-angle pattern provides information at atomic dimensions and short-range intermolecular repeat distances such as the intermolecular spacing of neighbouring phospholipids in the plane of the bilayer. By comparison with the preparation of specimens for electron microscopy, the proceedures involved in sample preparation for X-ray diffraction studies are simple, and it is possible to obtain diffraction patterns from freshly isolated and physiologically active nerve

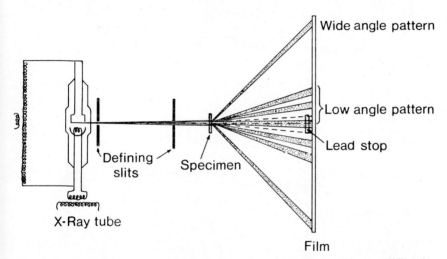

FIG. 4 (b). Arrangement of the main components in X-ray diffraction measurements.

bundles. Another structure which has been used for combined investigation using the techniques of electron microscopy and X-ray diffraction is the outer segment membrane of retinal rods, which has a highly ordered repetitious structure suitable for X-ray diffraction studies. From the X-ray diffraction patterns of these stacked membranes it is possible to determine directly the thickness of the repeating unit. By considering the relative intensities of the various orders of the diffraction pattern, an electron density profile through the membrane can be obtained. The problem of interpretation centres on the question of assignment of electron density peaks and troughs to structural features. As for electron microscopy, it is only possible to impose an interpretation based on what is previously known about membrane structure, the most significant feature of which is of course the lipid bilayer. What these approaches can do (and this is particularly true for X-ray diffraction) is to assign dimensions.

STACKED MEMBRANES: NERVE MYELIN

Before attempting to understand the details of molecular organisation in the repeating membrane structures of myelin or of retinal rod outer

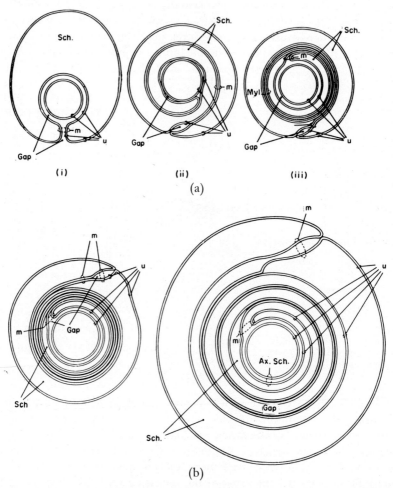

FIG. 5. (*a*) Diagram showing the stages in the development of peripheral nerve myelin: (i) the protofibre and the elementary mesaxon (*m*); (ii) intermediate stage; (iii) a myelinated fibre still at an early stage, showing the apposition of membranes along the mesaxon. (From ref. 9.) (*b*) The swelling of myelin in hypo-osmotic solutions separates the external membrane surfaces from each other, but the close apposition of the intracellular surfaces remains intact. The main accumulation of electron dense stain is at intracellular surfaces and this gives rise to the central dark line in the pentalaminar image of the double membrane. The external surfaces take up less stain and yield a less dense image. Reproduced by permission of Robertson (1960), *Prog. Biophys.* **10**, 343.

segments, on the basis of information derived from electron microscopy and X-ray diffraction, it is important to understand the gross morphology of the tissue under examination. The repeating structure of myelin is considered to arise as an extension of the plasma membrane of the Schwann cell plasma membrane, which, during development, encloses, and wraps around and around the nerve axon.[8] A schematic figure illustrating the process is shown in Fig. 5 and the appearance of myelin in the electron microscope is shown in Fig. 6. The gap (or mesaxon) marked "m" in Figs 5 (a) and (b) is a space, bounded by membrane on both sides, which connects the axon with the outside. The dense lines in the fixed preparation (also in the electron microscope photograph of a retinal rod outer segment, Fig. 6) arise from the close contact between the inner surfaces of the Schwann cell membrane, where all the cytoplasm appears to have been "squeezed" out. The less dense lines arise from deposition of the stain at the outer surfaces of the Schwann cell, in contact with the mesaxon, and the unstained zones arise from the hydrocarbon region of the membrane. The close apposition of two membranes gives rise to a pentalaminar arrangement, and this can be seen very clearly in the electron microscopic image of the individual discs of the retinal rod outer segment in Fig. 7.[17] Here, each disc has been flattened out by the exclusion of water by preparing the specimen in hyperosmotic conditions. The 180 Å repeating unit seen in these preparations corresponds to two membranes in very close apposition, and can be reconciled with the normal trilaminar image of other cell membranes on this basis.

The multilayered membrane structures of rod outer segments (Fig. 7) and myelin (Fig. 6) can also be looked upon as multi-water layers, and in the interpretation of the diffraction pattern, space has to be found for the water layers in the same way as it has to be found for the membrane layers themselves. This is the problem of "phase". The phase problem was basically solved when it was shown that the structure of the individual membrane layers was unaltered when the myelin is swollen in hypotonic solutions, but that the structure as a whole swells due to accumulation of water particularly in the intercellular space, i.e. the mesaxon.[19] Thus, the electron density profile due to features within the membrane should be invariant, but the profile due to features in the aqueous spaces should change with the extent of hydration of the specimen.

The electron density profile of rat sciatic nerve, determined from the first 5 order low angle X-ray reflections is shown in Fig. 8.[19] The most satisfactory interpretation of this electron density profile, which clearly supports the multilayered arrangement of myelin seen in electron microscope photographs, ascribes electron density troughs to the hydrocarbon regions.[20] The peaks are ascribed to protein and phospholipid head groups, where any heavy metal fixation would be expected to occur. The double-headed nature of the peaks arises from the apposition of two

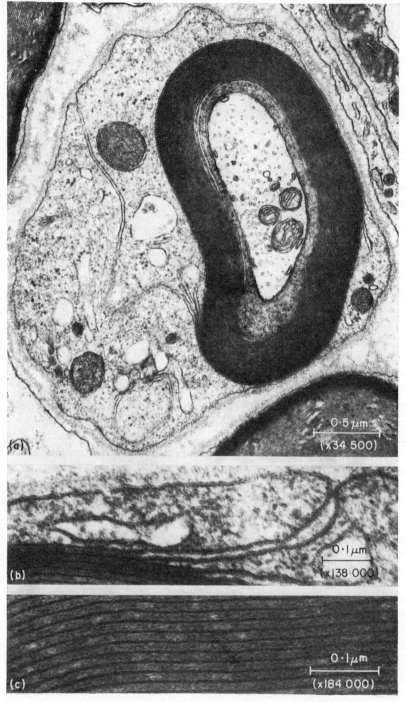

FIG. 6. *Caption at bottom of facing page.*

membranes. Recently, analysis of X-ray diffraction patterns of myelin up to the 18th order reflection have been reported, and these allow resolution of the electron density profile down to the 10 Å level.[21] At this degree of resolution, it begins to become possible to assign some details in the electron density profile to features in the molecular organisation of the membrane. In Fig. 9 the electron density profile of rat sciatic myelin is shown. It is essentially centrosymmetric but has some asymmetric features.

The interpretation clearly favours the lipid bilayer as the central structural fact of the myelin membrane. The electron dense peaks represent the lipid phosphate groups, which are the most dense components in the structure, and the width of these peaks is consistent with the dimensions of the polar head groups.

Absolute scaling of electron density indicates a density of 0·27–0.29 electrons Å^{-3} in the region of the trough and this is very close to the 0·27e Å^{-3} of long chain liquid paraffins and olefins. The terminal portions of the lipid hydrocarbon chains could on this basis, account for the measured electron density, and this sets an upper limit of 10% to the space occupied by protein in this innermost region of myelin. The asymmetric steps on either side of the trough are in the same positions as symmetric steps which are present in the density profiles of phospholipids prepared with cholesterol, but which are absent when the cholesterol is omitted. These steps then are assigned to cholesterol. Considerations of electron density in these regions indicate an approximately equimolar ratio of cholesterol and phospholipid on the external side of the bilayer, and a ratio of about 3:7 on the inner side.

The recent studies of other stacked membranes such as retinal rods provide proof of the structural bilayer in a limited number of systems.

Fig. 6. (a) Cross-section of a myelinated nerve fibre of the spinal root of an adult rat. The myelin is an extension of the plasma membrane of the Schwann cell, and it surrounds the nerve axon which can be seen at the centre. (b) Enlarged detail of (a), showing the gap at the entrance to the mesaxon, and the generation of the minor "intraperiod line" from the apposition of two extracellular surfaces in the myelin. (c) Enlarged detail of (a) illustrating periodicity, and the pentalaminar appearance of the closely apposed membranes of myelin. The major "intraperiod" line arises from the apposition of intracellular surfaces. In order to retain the structural integrity of the tissue, the preparation was fixed by intracardiac perfusion with a mixture of paraformaldehyde and glutaraldehyde.[16] This was chosen to obtain light but rapid initial fixation throughout the tissue by the readily diffusable paraformaldehyde, followed by the normal thorough lattice formation by cross-linking with the bifunctional glutaraldehyde. The sections were stained with lead citrate, followed by uranyl acetate. (Electron microscope photographs by D. Lawson.)

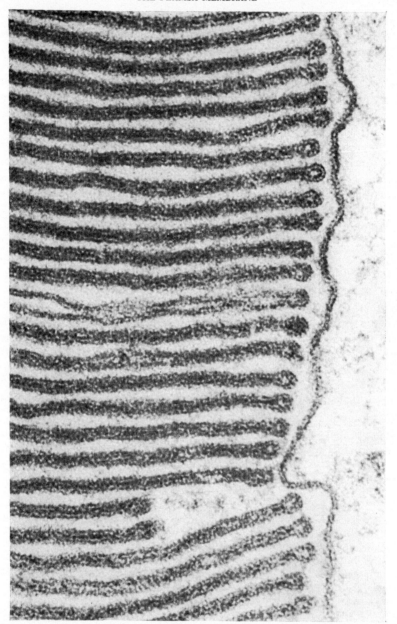

FIG. 7. Membranes of monkey retinal rod outer segments. This is a stacked membrane system, which has been exploited for combined electron microscope and X-ray diffraction measurements. The glutaraldehyde fixative was applied to the preparation in hyperosmotic solution, which caused most of the discs to collapse and obliterated the intra disc spaces. As a result, the discs appear mainly in the pentalaminar form of closely apposed double membranes. At the ends, where there is some separation of the disc membranes, the trilaminar image can be seen, and here it is indistinguishable from the plasma membrane. (From ref. 17.)

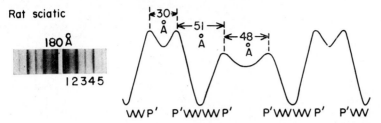

FIG. 8. Low-angle X-ray diffraction pattern of myelin showing the first five order reflections. The probable electron density profile is illustrated on the right of the figure, and the most likely positions of the heaviest atoms in the system (phosphorus) are indicated. The very low electron density trough is due to the methyl groups at the end of the phospholipid hydrocarbon chains. The shallow trough at the centre is due to the intracellular space and the two shallow troughs at the sides of the profile are due to the external water space of the mesaxon which connects the enclosed axon with the exterior. From Finean (1962), *Circulation*, **26**, 1151, by permission of the American Heart Association, Inc.

Some Comparative Aspects of Membrane Structure and Composition

The concept of the lipid bilayer as a general feature of membranes has been widely criticised on the basis that the best defined example (myelin) is also an exception among membranes. Thus, its function is thought to be only that of an electrical insulator for the axons which are enclosed within its folds. Unlike other nucleated cells, the Schwann cell (of which the myelin is the extension of the plasma membrane) does not appear to have surface receptors capable of translating external stimuli into signals for specialised cellular activity. In terms of composition too, it may be said to lie at an extreme of the range of variation, the cholesterol content being high, and the ratio of protein to lipid on the basis of weight, and on the basis of estimated areas occupied, being low (see Table 1). In fact, if membrane protein were spread in a simple monolayer as a β-pleated

TABLE 1. Protein and lipid composition of some animal and bacterial membranes[22]

	Molar ratio (cholesterol/phospholipid)	Area ratio (protein/lipid)
Myelin	0·7	0·43
Red blood cell	1	2·5
Bacillus licheniformis	0	4·8
Micrococcus lysodeikticus	0	4·3
Bacillus megaterium	0	5·4
Streptococcus faecalis	0	3·4
Mycoplasma laidlawii	0·11	4·1

N.B. The area ratios are calculated on the assumption that the protein is spread as a monomolecular film on the membrane surface, with each amino acid occupying 17 Å², each phospholipid 70 Å² and each cholesterol 38 Å².

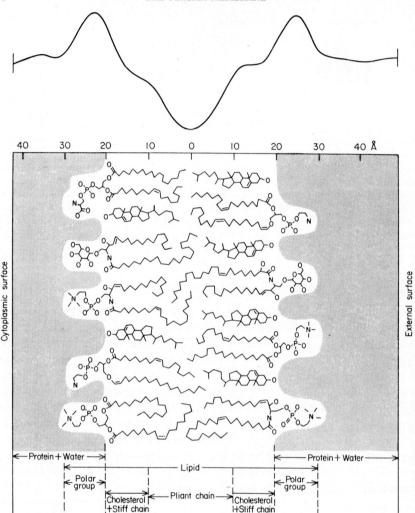

40   30   20   10   0   10   20   30   40 Å

Cytoplasmic surface   External surface

←— Protein + Water —→   ←— Protein + Water —→

|←——————————— Lipid ———————————→|

|← Polar →| |← Polar →|
| group | | group |

|Cholesterol|←— Pliant chain —→|Cholesterol|
|+Stiff chain| |+Stiff chain|

FIG. 9. Electron density profile through rat sciatic myelin, based on the first 18 order reflections. The regions of high electron density are associated with the phosphate groups, and the region of lowest electron density is associated with the region of interface between the terminal methyl groups. The shoulders on the peaks are due to cholesterol, and are absent from electron density profiles of membrane preparations lacking this component. The information provided by X-ray measurements enables a fairly precise plan of molecular organisation in membranes to be built up. Reproduced by permission of Caspar and Kirschner (1971), *Nature New Biol.* **231**, 46.

sheet on the membrane surface (as was originally proposed for the "unit membrane" structure), then only in myelin would it be possible for the area occupied by lipid to exceed that occupied by the protein.[22] In all other membranes the area occupied by protein would be greater than the lipid, and extensive regions of the membrane would have to be composed of protein only.

The variation in the composition of membranes within a single tissue is also wide: thus the weight ratio of phosphorus to lipid in beef liver mitochondria is 0.033, but 0.019 in the plasma membrane of the same organ. This difference is explained by the large amount of cholesterol present in the plasma membrane, and its absence from the mitochondrial inner membrane. Similarly, cardiolipin (diphosphatidyl glycerol) is the chief acidic phospholipid of mitochondria, but in mammalian tissues this is probably its exclusive location. The functional roles and the dimensions of the other membranes within tissues and individual cells also vary widely.

THE UBIQUITOUS BILAYER

Can it be said then, that the bilayer, probably proven for myelin and supported by much circumstantial evidence for other membranes, is the universal structural feature which defines "membrane"? Recent developments have allowed X-ray diffraction techniques to be applied to dispersions of phospholipids and membranes in suspension, and whilst the measurements are not yet sufficiently refined to allow the calculation of the electron density profile along the line normal to the plane of the membrane, and hence the definition of molecular organisation in this dimension, they do provide strong evidence in favour of the bilayer concept. This is best exemplified by the membranes of *Acholeplasma laidlawii*[23, 24] (formerly *Mycoplasma laidlawii*). As was described in Chapter 1, the single membrane of this simple organism has been of particular interest in the field of membrane science because fatty acids provided in the growth medium are incorporated unchanged into the hydrocarbon component of the membrane phospholipids and in this way it is possible to control such features as hydrocarbon chain length and saturation.[25] Sterols may be incorporated into the membrane in the range 0–12% of total lipid.[26] The X-ray diffraction patterns[24] from *Acholeplasma* membrane dispersions at low temperatures (10°) indicate that methyl groups tend to be localised in a region at the centre of a bimolecular structure, and that the thickness of this structure agrees with that calculated for the thickness of a membrane composed of lipids lying fully extended in a direction perpendicular to the plane of the membrane. Thus, the thickness of the bilayer in *A. laidlawii*, in which the predominant hydrocarbon is 22:1-cis, is greater than that in which the predominant hydrocarbon is 16:0. On raising the temperature to 40°, the evidence for an ordered region of methyl groups is absent and the membrane thickness decreases as the hydrocarbons become less formally extended. Similar diffraction patterns have been obtained from suspensions of red cell membranes and nerve endings.[23]

The bilayer concept, which was so strongly supported by the early attempts at high resolution electron microscopy, ran into problems of interpretation, particularly with regard to the introduction of artefacts due to the chemistry of fixation. The problems facing X-ray diffraction studies are mainly those of interpretation: sample preparation for X-ray study is unlikely to introduce major artefacts, though one needs to be sure that the X-ray beam itself does not interfere with the specimen under investigation. But there is a critical barrier which prevents most biologists (and chemists too) from proceeding beyond the fundamental Bragg equation to the level of comprehension required to assign molecular organisation in membranes from X-ray diffraction evidence. The proof of the bilayer by X-ray diffraction then has to be taken on trust from its practitioners but, quite plainly, there is much more that remains to be said about membrane structure. For the remainder of this chapter, our concern will be mainly with the placement of protein within and about the central phospholipid bilayer. The simple "unit membrane" hypothesis suggested that the protein might be spread asymmetrically at the two aqueous interfaces of the bilayer, in a β-extended configuration, but against this we have seen that for most membranes (having a high ratio of protein to lipid) this kind of organisation becomes an impossibility as soon as we accept the idea that it is the phospholipid bilayer continuum which sets the limit on the surface area of the membrane. We must therefore consider the possibility that membrane protein might exist in compact rather than extended configuration, and that it may have locations other than at the surface.

FREEZE-FRACTURE ELECTRON MICROSCOPY[27]

A most attractive development in the field of membrane science has been that of freeze fracture and freeze-etch electron microscopy. Freeze fracture techniques share with all other form of microscopy a vital advantage over all other forms of physical investigation: a wide field, arrested in time, can be examined down to the limits of resolution. Questions can be asked about the significance of features which may occupy 1% or less of the total field. In contrast, all other physical techniques applied to membrane structural determination are both time- and spacially-averaged. Thus, minor components and, even more importantly, transient structures and arrangements are generally averaged out as "noise". Furthermore, freeze fracture and freeze-etch electron microscopy can avoid some of the problems of chemically induced fixation artefacts which beset the attempts at defining membrane structure by thin-section techniques. Most important of all, this form of sample preparation allows an entirely new dimension of membrane organisation to be examined.

Sample Preparation

The basic steps in the preparation of a sample for freeze fracture and etching are illustrated in Fig. 10.[27] The tissue sample (dimensions about 1 mm³) is rapidly cooled to the temperature of liquid nitrogen ($-196°$).

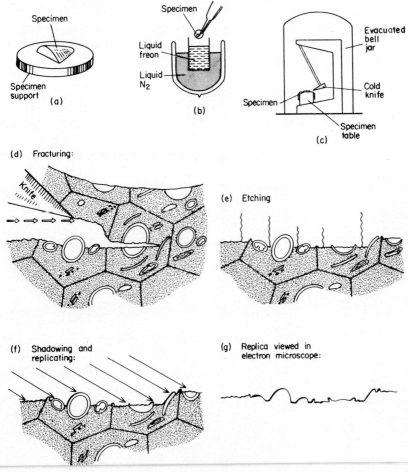

FIG. 10. Freeze etching. The specimen (a) is placed on a copper disc and (b) rapidly frozen by immersion in liquid Freon 22, cooled to the temperature of liquid nitrogen. The frozen specimen is then placed in the precooled vacuum chamber (c) and fractured (d) with a cooled microtome knife. In some cases, the fractured specimen may be etched (e) by continued exposure to high vacuum. The surface is shadowed and replicated with platinum and carbon (f) and after dissolving away the specimen, the replica (g) can be examined in the electron microscope. (From ref. 27.)

The frozen material is then mounted on a refrigerated support in a high vacuum chamber (2.10^{-6} torr) and fractured with a cold microtome knife. At this point, the newly revealed surface may be shadowed and replicated by vacuum deposition from platinum and carbon sources placed in the chamber so that the electron dense shadowing material falls obliquely on to the fractured specimen. Alternatively, the specimen is kept under high vacuum for a few seconds after fracturing and before shadowing, to allow ice on the fracture surface to sublime. This "etching" process lowers the ice table by about 1000 Å, but has no effect on the surfaces of the biological material. The freeze-etched preparation is then shadowed and replicated as before, and the replica can be mounted and examined directly in the electron microscope.

Images and Interpretation

A freeze-fracture (non-etched) preparation of a red cell membrane is illustrated in Fig. 11.[28] The membrane is partially submerged in a sea of ice. The fracture surfaces revealed by the process are coated with particles of about 85 Å diameter: the surface designated PF in Fig. 11 has a higher

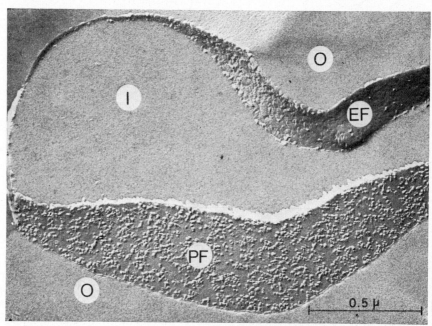

FIG. 11. The appearance of the fracture planes of a red cell membrane. An explanation of the symbols is presented in Fig. 13; the outward facing protoplasmic surface is heavily studded with intercalated (or intramembraneous) particles; the particle density on the inward facing, external face F is much lower. (From ref. 28.) Figs. 12, 13, 14, 15 reproduced by permission of Pinto da Silva and Branton (1970), *J. Cell. Biol.* **45**, 598.

density of these particles than the surface EF. (In this account the fracture and etch surfaces are designated according to a new proposal[29] emanating from a group of practitioners in the field of freeze-fracture electron microscopy, which, it is to be hoped, may supersede the bewildering assortment of nomenclature which is current in the literature.) Figure 12 shows an

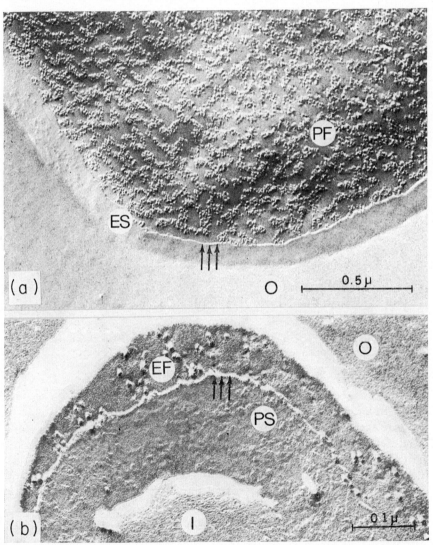

FIG. 12. The appearance of the planes in a preparation which has been fractured and etched. A ridge (triple arrow) characteristically separates the fracture faces from the etch faces. (a) A convex fracture revealing the protoplasmic (PF) face and external etch (ES) face. (b) A concave fracture revealing the external (inward facing) EF fracture face and the internal PS etch face. (From ref. 28.)

electron microscope photograph of a preparation which has been both freeze-fractured and etched; in freeze-etch preparations we see new surfaces PS and ES which are clearly separated from the fracture surfaces PF and EF by a ridge. There is no sign of the 85 Å particles on the etch surface and by comparison with the fracture surfaces the new surfaces revealed by etching are rather smooth: there is no sign of the 85 Å particles. There is much evidence to support the identification of the fracture and etch surfaces according to the scheme[28] in Fig. 13. The heavily studded fracture faces are not true membrane surfaces, but are thought to be derived from the splitting of the membrane in the plane of the hydrophobic layer, like dismantling a sandwich. The fracture face with the

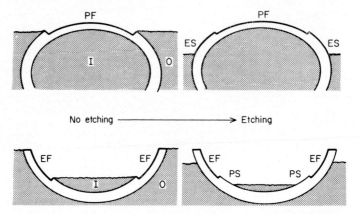

FIG. 13. Rationale of the fracturing and etching processes. Without etching, fracturing exposes convex (F) and concave (F) fracture faces. When the newly revealed surface in exposed to high vacuum, the ice inside (I) and outside (O) is lowered, exposing a convex (E) and a concave (E) etch face. (From ref. 28.)

denser population of 85 Å particles (PF) is differentiated from the other fracture face (EF) by the recognition that (PF) is outward facing and coherent with the protoplasmic, (P) half of the sandwich, whilst the (EF) fracture face is inward facing and coherant with the extracellular (E) half sandwich. The smooth etch faces are considered to be true membrane surfaces (S), and an examination of their spacial relationship with the ice planes makes the identification of the extracellular surface (ES) and protoplasmic surface (PS) a relatively simple matter. The ridge which is seen in etched preparations represents the portion of the membrane thickness which separates the fracture and etch planes. Under favourable circumstances, the PF and ES planes are clearly convex, while the EF and PS planes are concave in shape, and this lends credence to the designation of the surfaces indicated in Fig. 13.

   This interpretation has been supported by freeze fracture studies of stearic acid bilayers in which the site of fracture lies in the plane of the inner methyl end groups of the hydrocarbon chains.[30] It seems likely that a frozen membrane would fracture in a similar zone. An alternative approach has been to show that crosslinking of proteins by prior fixation with the bifunctional glutaraldehyde does not alter the electron microscopic appearance of the freeze-etch specimen (in spite of a demonstrable improvement in their mechanical stability).[31] The effect of lipid extraction or fixation with osmic acid[32, 33] prior to freezing is however drastic, with the effect of removing all the fracture planes. This is quite contrary to the effect of defatting prior to fixation in the preparation of thin sections for electron microscopy[13, 31] (see p. 56) in which the appearance of the final image is preserved. Thus it is seen that the presence of lipid in membranes is a requirement for freeze-cleavage to be possible.

The etch surfaces have been identified positively by the attachment of labels which can be recognised in the final electron microscopic image.[28] The red cell membranes which appear in Figs 11 and 12 were prepared in the presence of a high concentration of ferritin. This protein has a dense iron-containing core, and with a diameter of 120 Å is easily identified in electron microscope images when present; no conjugating agent was used to couple the ferritin covalently to the red cell membranes, and nothing resembling ferritin can be observed. When the membranes were treated with ferritin in the presence of the bifunctional amino-reactive reagent toluene-2,4-diisocyanate, the iron-protein bound to both the extracellular and the intracellular surfaces, as was shown by conventional thin section electron microscopy (Fig. 14). Freeze-etch electron microscope images of

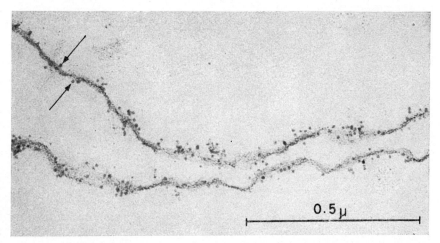

0.5 μ

FIG. 14. Thin-section electron microscope photograph demonstrating that covalently bound ferritin is associated with both surfaces of the membrane. Fixed with glutaraldehyde, postfixed with osmium tetroxide, stained with uranyl acetate, and poststained (after embedding and sectioning) with lead acetate. (From ref. 28.)

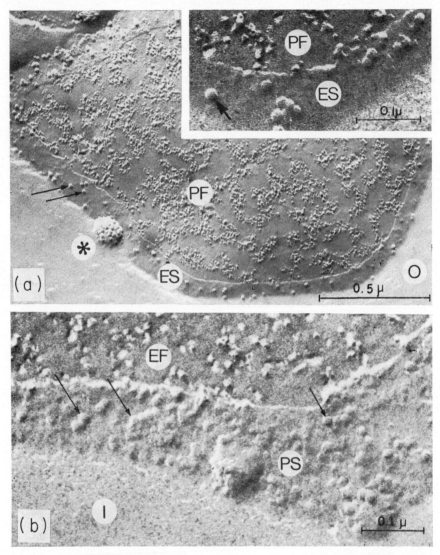

FIG. 15. Etched and fracture faces of a ferritin-conjugated red cell membrane. The ferritin molecules (arrows) are associated exclusively with the etched faces. These photographs may be compared with Fig. 12. which was prepared from red cell membranes in the presence of ferritin, but omitting the bifunctional covalent coupling agent, toluene-2,4-diisocyanate. The globular structure (*) may be artefactual. (*a*) Convex fracture revealing the protoplasmic PF fracture face and the external surface ES of the membrane; (*b*) concave fracture revealing the external fracture face EF and the internal etch surface PS of the membrane. (From ref. 28.)

FIG. 16. (a) Appearance of the convex fracture plane of a red cell membrane which has been treated with pronase to remove 70% of the total protein. (b) The freeze-etch image of a control cell which has been incubated in buffer only, and retains its original protein. Treatment with pronase also results in the loss of the membrane intercalated particles, and on this observation was based the proposal that they are composed of protein. (a) from Engstrom; Thesis, University of California, Berkeley 1970; (b) from Daniel Branton.

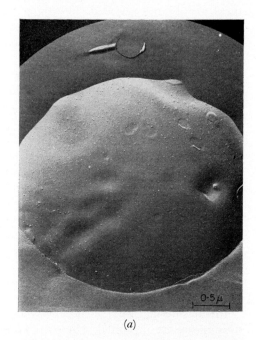

(a)

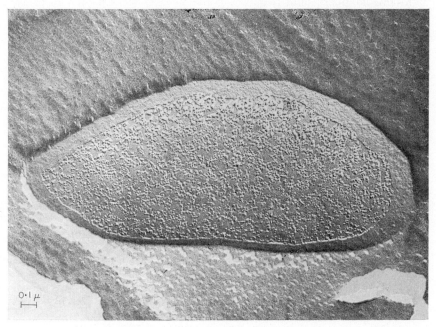

(b)

these membranes revealed the conjugated ferritin, bound exclusively to the etch faces PS and ES. In a related study the membranes of isolated sea urchin nuclei were examined.[34] These have an "intrinsic" label – ribosomes bound to the membrane surface. The freeze-etch image of the nuclear membrane revealed particles of the size of ribosomes exclusively located on the etch faces, confirming the idea that the etch faces are derived from the original surfaces of the membrane.

Identification of 85 Å Intercalated Particles

What can be said of the numerous small particles which are found only within the hydrophobic region, and which are revealed by the fracture process? Generally, it may be said that membranes having a high biological activity provide fracture surfaces with a high density of particles.[27] Thus, the highest particle densities are observed on the fracture faces of chloroplast lamellae[32] (with 80% of the fracture surface area covered by particles) and retinal rods,[35] while no particles have been observed on the fracture faces of myelin membranes[36] and phosphatidyl choline liposomes.[37] As in the case of the red cell membrane, the distribution of particles between the two fracture faces is generally unequal.

In considering what structures these particles represent, two main possibilities are suggested. The particles might be derived from patches of non-lamellar phase lipid (e.g. hexagonal phases, see Chapter 1, p. 13). Alternatively, they could represent protein, possibly active enzyme components of membrane function. Freeze-fracture of a variety of different lamellar phase lipids, having differing chain length, degree of unsaturation, polar groups, etc., always reveals an extended smooth planar image.[37] Hexagonal phase lipids produce typically different images having ribbed patterns and at least two fracture planes at an angle of 120° to each other. The characteristic freeze-fracture images of lipid phases are distinct and recognisable, and quite clearly different in appearance from the freeze-fracture face of biologically active membranes.

FIG. 17. (a) Convex freeze-etch image of a blood group A human red cell membrane. The membrane was treated with ferritin labelled anti-A serum, and by manipulations designed to aggregate the intercalated particles of the fracture plane. The aggregated particles are also visible as roughened zones on the etched PS surface of the membrane, and a continuity of these roughened zones with the particle aggregates across the ridge separating the two planes can be discerned. In (b) the particle aggregates and the roughened zones have been outlined in order to make this clearer. The ferritin molecules (circled in (b)) are located exclusively in the roughened zones on the PS etch face so that they override the aggregated particles in the fracture plane. No ferritin is to be seen on the fracture face, nor on the smooth regions of the etch surface. For these reasons it was concluded that protrusions of the membrane intercalated particles are expressed through to the outer surface of the membrane. (From ref. 39.) Reproduced by permission of Pinto da Silva et al. (1971), Nature New Biol. 231, 46.

It is now universally agreed that the intercalated particles of the fracture face are composed of protein, and this is certainly supported by the finding that treatment of red cell membranes with the proteolytic enzyme, pronase, results in a fracture image which is devoid of particles.[27]

The intercalated particles of the red cell membrane appear to be

(a)

(b)

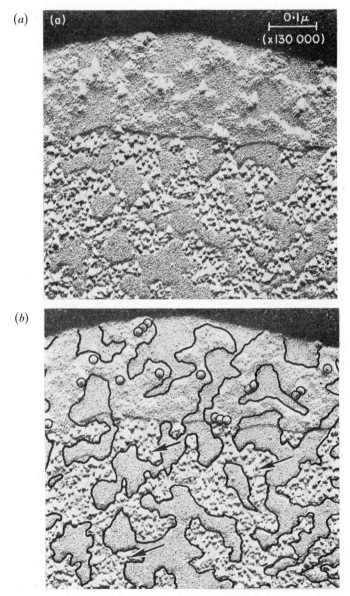

FIG. 17. *Caption at bottom of facing page.*

potentially mobile.[38] By simple manipulations, such as reduction of temperature, ionic strength or pH, one may cause the particles of the fracture face of isolated red cell membranes (but not the membranes of intact red blood cells) to coalesce into discrete *aggregates* within a minute or two, leaving smooth areas quite devoid of particles. The aggregation phenomenon is reversible and by readjustment of the conditions the aggregates can be redispersed as monomers in the fracture plane. Aggregation of the particles can be prevented by prior fixation with the bifunctional amino group reagent glutaraldehyde, and this is further evidence for their proteic nature. The aggregation process must be mediated by translational movement of the particles within the plane of the membrane, which must now be envisaged as a planar fluid domain formed by the phospholipid bilayer, interrupted by localised yet mobile protein inclusions.

If the particles are aggregated by treatment of the membranes with distilled water prior to the freeze-fracture operation, then roughened disturbances are revealed on the normally smooth etch surface, which show continuity with the particle aggregates of the fracture plane[39] (Fig. 17). When particle aggregated membranes derived from an A-group blood donor were reacted with ferritin-labelled antibodies to blood group A, the ferritin marker was found to be bound almost exclusively in the roughened regions of the etch surface E, contiguous with the internal aggregates. Virtually no ferritin was bound to the smooth areas. Thus, although the particles are mainly associated with the cytoplasmic half of the membrane after fracturing, they extend through the outer half and are expressed as antigenic determinants at the membrane surface.

The freeze-fracture technique begins to yield detailed information about the microscopic topography of the membrane. The evidence for bilayer structure is, if anything, strengthened by this new approach, but we begin to see details included within this general structure which are not readily visible to techniques which are dependent on time or spatial averaging (spectroscopy, diffraction and calorimetric techniques) and which are lost in thin-section electron microscopy due to problems of preparational artefact, resolution and interpretation. Most importantly, we can expect to identify and locate certain functional loci, which are certainly invisible to all other forms of examination.

MEMBRANE PROTEINS

The architecture of membranes has so far been considered almost exclusively in terms of the molecular interactions of lipids. Whilst recognising that lipids alone may define the structural framework of biological membranes, it is obvious that the interesting functional properties of membranes are derived from the association of specific enzyme proteins with the special environment provided by the lipid matrix. The conformation of proteins, and hence the reactivity of enzymes, is highly sensitive to environment. The main determinant of environment that we should consider, when comparing the behaviour of proteins in aqueous solution and in membranes, is that of polarity. The early models of membrane structures ascribed mainly ionic forces to account for the association of proteins with lipids.[6, 7, 8] The unit membrane hypothesis, with protein in an extended conformation spread over the polar surface plane of the lipid bilayer was the ultimate expression of this idea. We should be wary of this concept. Even in aqueous solution, non-ionic forces have been shown to contribute significantly to the stability of multi-subunit protein structures: the binding of $\alpha\beta$ haemoglobin dimers to form the $\alpha_2\beta_2$ tetramer is a fine example of this.[40]

With the hydrophobic quality of the membrane interior in mind we might expect forces other than those of a purely ionic character to be of importance. It is well known that manipulations which normally disrupt ionic linkages, such as alterations of pH and ionic strength, are not effective in releasing all the protein from membranes into solution.[41] By way of contrast, and as a pointer to the nature of protein-lipid interaction, regular techniques of membrane protein solubilisation include treatment of membranes with solvents (n-butanol[42]), denaturants[43] and detergents.[44] The efficacy of such procedures is strong evidence for hydrophobic forms of association between the components of membranes.

As in the case of membrane lipids, there are two general approaches which can be undertaken to try and understand the disposition of proteins in membranes. We can try and purify the component species following general or selective extraction of proteins from the membranes. A programme could be set up which might include detailed structural examination of the isolated protein, together with a study of conformation and enzyme activity in water and organic media, leading possibly to incorporation of the protein in model membrane systems in order to understand structure-activity relationships of a membrane protein. Some investigations along these lines have been instigated, and these include such key membrane activities as the synaptic acetylcholine receptor site,[45] the calcium-dependent ATPase of sarcoplasmic reticulum[46, 47] and the anion exchange pathway of the red cell membrane.[48] This approach, which will ultimately be the more potent, leads to the solution of specific problems concerning biological activity in membranes. The other approach, generally by the spectroscopic study of membranes in suspension, leads to general (space and time averaged) ideas about protein disposition in membranes.

Protein Spectroscopy
A detailed description of the properties of membrane proteins is well

beyond the scope of this account, and so a brief summary of some of the conclusions which may be drawn from this approach must suffice.[49] The evidence of circular dichroic (CD) and optical rotary dispersion (ORD) spectra (closely related manifestations of optical activity) and of infra-red spectroscopy, rules against the presence of an extensive contribution of protein in β-extended conformation. Whilst globular structures seem to be the rule, it is difficult to assess from the spectroscopic approach how large is the contribution from helical conformations and how much should be termed non-symmetrical or random. Certain regions in the CD and ORD spectra of model polypeptides are highly sensitive to the conformation of the secondary structure and under ideal conditions it is sometimes possible to use this spectroscopic approach to assign the contribution of α-helical, β-extended and non-symmetrical (random coil) conformation. Figure 18 shows the ORD spectrum of a suspension of plasma membranes from

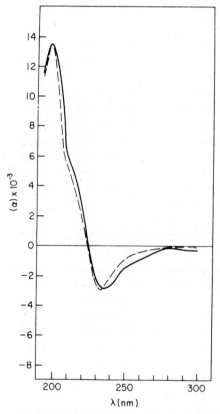

FIG. 18. Optical rotatory dispersion spectrum of the plasma membrane of Erlich ascites tumour cells in aqueous suspension. ————, plasma membrane; - - - -, poly-L-glutamic acid (an α-helical structure). The magnitude of the spectrum of the poly-L-glutamic acid has been reduced by a factor of 4·5. (From ref. 50.)

Erlich ascites tumour cells, together with a spectrum of α-helical poly-L-glutamic acid.[50] At first some correspondence between the two spectra is apparent, but there are a number of anomalies which make spectral interpretation exceedingly difficult in terms of the well-recognized ideal conformations which are known to exist in proteins. Generally the form of the spectrum is similar to that given by the α-helical reference substance, but the amplitude of the spectrum should be greater (the spectrum of the poly-L-glutamic acid has been reduced by a factor of 4·5), and the spectrum of the membrane preparation is somewhat red-shifted.[49] It has been argued that spectral contributions from conformations other than α-helical are being quenched, allowing us to see only the α-helical component, but if this were the case, then the spectrum would be blue-shifted (i.e. to the left of the reference substance).

Without a very secure understanding of the physical basis of the spectral anomalies which could lead to the production of corrected spectra, these data cannot be used to provide the precise information which would be required to assign accurately the conformations of the membrane protein. Attention has been focused recently on light scattering and absorption artefacts to which these spectra are particularly sensitive,[51] and also to a reconsideration of reference molecules.[51a] The ORD and CD spectra of water soluble globular proteins are also subject to anomalies not dissimilar to those encountered when examining membranes, and it is likely that these arise from the apolar, highly polarisable environment in which most of the peptide linkages are placed. In spite of the precision with which the conformation of reference substances is known, their relevance to proteins in general, and membranes in particular, is questionable, due to the extensive degree of hydration throughout these molecules. There are many other grounds on which the use of simple model polypeptides can be criticised when used as spectral reference compounds for membrane proteins,[51a] but in spite of these complications (which apply equally to globular and to membrane proteins) it is widely agreed that the spectral evidence indicates a very significant contribution of α-helical conformation in biological membranes (greater than 50% in the case of red cell membranes) and that membranes which are particularly active in terms of ion transport have a smaller proportion of protein in this particular conformation (and possibly a larger proportion in β-conformation). The CD and ORD spectra of membrane proteins are very sensitive to manipulations which perturb phospholipid,[50, 51b] such as treatment with digitonin, lysolecithin and enzymes (phospholipases A and C), and this is indicative of an intimate relationship between these two major components of membrane structure in the hydrophobic environment of the membrane interior.[50] Despite the inability to assign precise conformations, it is clear that this means of investigation dealt the death-blow to the earlier notions exemplified by the "unit membrane" hypothesis, in which the protein, spread as a β-pleated sheet, was located at the membrane surfaces, and held there by electrostatic forces. From here on it was essential to consider protein-lipid interactions in membranes in terms of hydrophobic forces, similar to those responsible for the stabilisation of the phospholipid molecules in the bilayer itself.

It was from their experience with the spectroscopic methods (mainly CD, ORD, IR, and fluorescence) that Wallach and Zahler[50] were able to propose that membrane proteins have unique structures (distinct from the water-soluble globular proteins) in which two hydrophilic peptide regions are widely separated from each other by a hydrophobic zone. The hydrophobic unit would have a dimension equal to the hydrocarbon core of the membrane phospholipid bilayer, and the hydrophilic sections would be exposed at the two membrane surfaces. In general, the hydrophobic units would consist of helical peptide segments, and by aggregation it could be possible (depending on the primary sequence of amino acids) to form a channel through the membrane which could have a polar interior. Transport of ions and other hydrophillic solutes could occur through such structures, which could be subject to control either by varying the state of subunit aggregation, or by regulating the conformational state of the protein monomers. At the surface of the membrane, the hydrophilic peptide sections would be able to interact with the exposed sections of other transmembrane proteins, and with proteins dissolved in the aqueous phase.

Subsequent experience with membrane proteins has borne out the truth of some of these predictions, and it will be useful to bear them in mind when reading the remainder of this chapter, and the latter part (pp. 139–174) of Chapter 3.

Resolution of Membrane Proteins

Clearly, the instrumental approach to understanding the integrated functions of membranes is at its most potent when applied to highly specialised systems having only a single protein: as examples of these we have rhodopsin, which comprises 85% of the protein in cattle rod outer segment membranes,[52] and bacteriorhodopsin, which is the unique protein of the purple membranes of *Halobacterium halobium*.[53] Thus it has been possible to show, by specialised electron microscopic techniques, that bacteriorhodopsin is a globular protein, with about 70–80% of its residues organised in α-helical conformation, and that it almost certainly traverses the membrane bilayer to be expressed at either surface.[54] Nearly all other membranes contain a multiplicity of proteins and the value of spectral and other instrumental information related to individual functions is reduced in the face of a massive "background" contribution due to other proteins.

It is obvious that membrane proteins must be studied individually. In order to obtain the best resolution of proteins, the membranes are normally

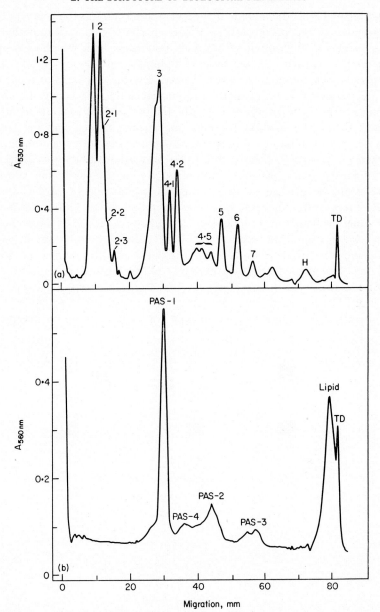

FIG. 19. Densitometer traces of polypeptides (*a*) and glycoproteins (*b*) of red cell membranes after resolution by SDS-polyacrylamide electrophoresis. Each gel was loaded with 40×10^{-6}g total protein and were stained with Coomassie-blue for protein and with the periodic acid-Schiff reagent for carbohydrate. H indicates the position of polypeptide chains from haemoglobin which are always present as a contaminant in red cell membrane preparations, and TD indicates the position of the low molecular weight anionic tracking dye which is added to the protein mixture and used to follow the course of the electrophoretic separation while it is in progress. (From ref. 55.)

disaggregated with the anionic detergent, sodium dodecyl sulphate (SDS), (sometimes in the presence of a reducing agent, e.g. mercaptoethanol). This releases all the proteins from their association with the phospholipid. The mixture is then resolved by electrophoresis on polyacrylamide gel in the presence of the same reagents. Essentially this is a form of molecular sieving, because the charge on the proteins treated in this way is due to the anionic centre of the detergent molecule, and the proteins separate according to size. Thus, the smallest proteins migrate the fastest, and the largest slowest, and by suitable calibration and comparison with macromolecular markers of known size the method can be used to determine molecular size directly. After the electrophoresis, the positions of the resolved proteins in the gel can be determined by application of various stains: amido-black and Coomassie-blue are commonly used. Other features of individual proteins can sometimes be identified by the use of more specific stains; for example, the application of the periodic-Schiff reagent, which is specific for carbohydrate (particularly sialic acid), identifies the glycoproteins. The stained gels are commonly photographed, sketched or they may be scanned in a densitometer. In Fig. 19 (*a*) and (*b*) the densitometer scans of SDS polyacrylamide gels containing the resolved proteins of red cell membranes stained for protein and for carbohydrate are illustrated,

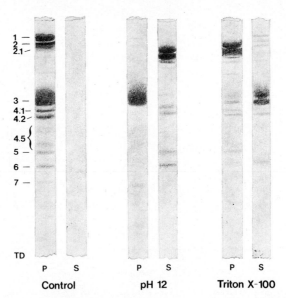

FIG. 19. (*c*) Selective solublisation of polypeptides from red cell membranes by exposure to alkaline conditions (pH 12) or to the non-ionic detergent Triton X-100. After treatment and centrifugation, samples of the insoluble pellet (P) and of the supernatent (S) equivalent to equal quantities of original membranes were taken up in SDS and electrophoresed in polyacrylamide gels. The resolved proteins were stained with Coomassie-blue. (From refs. 43, 44.)

where they are numbered according to a widely used convention.[55, 78] A number of the electrophoretic bands have been identified and associated with specific properties and functions, and these are listed in Table 2.[55]

Some of the proteins associated with the membrane may be detached by relatively mild proceedures which do not totally destroy the integrity of the membrane organisation, and Fig. 19 (c) illustrates photographs of SDS gels of proteins eluted from the membranes, and retained by the membranes under various conditions.[43, 44, 55] It can be seen that exposure of the membranes to alkaline media (pH 12) releases bands 1 and 2 ("spectrin", see Table 2), an actin-like polypeptide (band 5) and the glycolytic enzyme glyceraldehyde-3-phosphate dehydrogenase (band 6). The non-ionic detergent Triton X-100 releases other proteins from the membranes, most notably the broad and intensly staining band 3; other proteins remain firmly attached to the insoluble material and can only be soublised with denaturing detergents such as SDS.

TABLE 2. The major polypeptides and glycoproteins of red cell membranes[55]

Component	mol. wt.	Percentage of stained protein[a]	Polypeptide chains/membrane Ghost[b]	Characterisation, and other designations
A. Polypeptides				
1	240,000	15·1	216,000	⎧ Spectrin
2	215,000	14·7[c]	235,000[b]	⎨ Tektin A
				⎩ Myosin-like polypeptides
3	88,000[d]	24·1	940,000	Component a
				Protein E
				Minor glycoprotein
4.1	78,000	4·2	180,000	
4.2	72,000	5·0	238,000	
5	43,000	4·5	359,000	Actin-like polypeptide
6	35,000	5·5	540,000	G3PD
7	29,000	3·4	403,000	
B. Glycoproteins				
PAS-1 ⎫	55,000	6·7[e] (2·8)	500,000	⎧ Glycophorin
PAS-2 ⎭				⎨ Sialoglycoprotein
				⎩ ABO and MN isoantigen

[a] Average values from a laboratory exercise in which 21 medical students determined the distribution of Coomassie-blue stain on gels of their own red cell membranes, by the method of Fairbanks et al.[78]

[b] Calculated assuming that each haemoglobin-free red cell membrane contains 57×10^{-12} g protein.[78]

[c] Possibly subject to a slight overestimation due to the overlap with band 2·1 (see Fig. 19a).

[d] This is the value for the peak of the stained band; the tail extends out to 105,000.

[e] Calculated with the assumption that this glycoprotein carries 70% of the total membrane sialic acid and that 1·6% of the membrane mass is sialic acid. The figure in parentheses is for the protein portion alone.

In general, electrophoresis of membrane proteins on polyacrylamide gel must be considered to be a terminal technique; a normal "load" for electrophoresis is small (about 10^{-5} g) and the condition of the resolved proteins after treatment with SDS is such that further experimental work at a functional level is not generally possible. Sodium dodecyl sulphate is a potent denaturant. In spite of these obstacles, a great deal of imaginative work has been done with the simple resources which are needed to run polyacrylamide gels, and this has yielded much valuable information.

MEMBRANE ASYMMETRY

The techniques outlined above are obviously ideal for the study of genetic and pathological variation in membrane proteins and, more ambitiously, attempts have been made to study the asymmetry properties, or sidedness of the red cell membrane. The general approach is to treat the red cells with a marker reagent to which the membrane is impermeable, and which attaches covalently to selected groups on the membrane. Usual marker properties are radioactivity and fluorescence. Alternatively, the membrane surface can be modified by gentle enzyme catalysed hydrolysis. Because the membrane is impermeable to the marker reagent or enzyme, only reactive groups located on the outer surface of the membrane are able to react. The membranes are then isolated by osmotic haemolysis and centrifugation, solubilised with detergents, and the proteins separated by electrophoresis.

TABLE 3. Labelling reagents for the membrane outer surface

Reagent	Probable target groups	Label	Functional effects
A 4-acetamido, 4′-isothiocyano stilbene-2,2′-disulphonate (SITS)[58]	amino, histidyl, guanidyl	fluorescence	Inhibits anion exchange[62]

B sulphanilic acid diazonium salt[59]	tyrosine	^{35}S	Inhibits acetyl-cholin-esterase[60] Increased permeability[59] to Na^+ and K^+ Inhibition of glucose transport

TABLE 3—*continued*

C

formylmethionyl sulphone   lysine ^{35}S
  methyl phosphate [57]
  (FMMP)

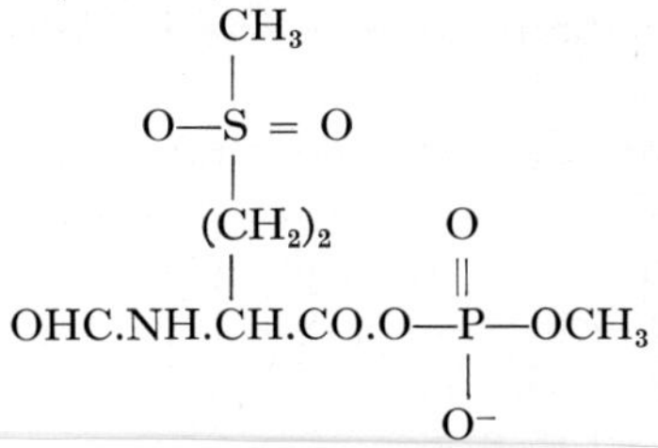

$$CH_3$$
$$|$$
$$O—S = O$$
$$|$$
$$(CH_2)_2 \qquad O$$
$$| \qquad\qquad ||$$
$$OHC.NH.CH.CO.O—P—OCH_3$$
$$|$$
$$O^-$$

D

trinitrobenzene sulphonate[64]   amino, sulphydryl ^3H Inhibits anion
  (TNBS)   exchange[62]
                                                                      Increases cation
                                                                        impermeability

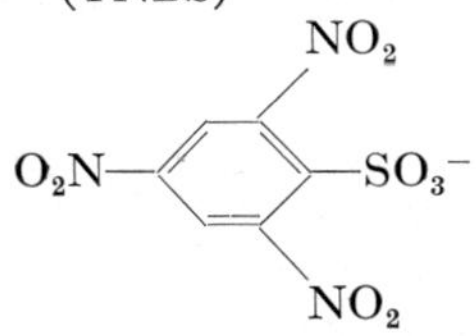

$$NO_2$$

$$O_2N—⟨\ ⟩—SO_3^-$$

$$NO_2$$

E

lactoperoxidase catalysed   tyrosine, ^{125}I
  iodination[61]   histidine
protein $+ H_2O_2 + I^-$

F

pronase catalysed   altered   Inhibits anion
  proteolysis[60, 63]   proteins   permeability[62]
                                                                Increases cation
                                                                  permeability[65]

G

N-(4-nitreno-2-   entirely non- ^3H or ^{35}S
  nitrophenyl)-2-aminoethyl   specific adduct
  sulphonate[66, 67]   to C-H bonds
produced by photolysis of   of protein, car-
  azide precursor   bohydrate and

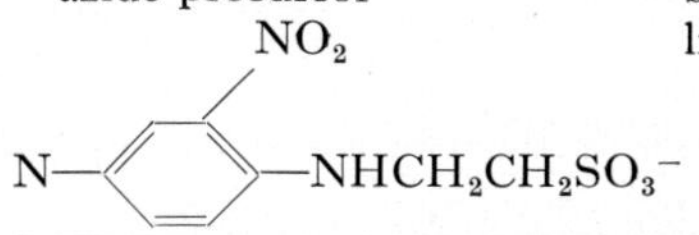

$$NO_2$$ lipid

$$N—⟨\ ⟩—NHCH_2CH_2SO_3^-$$

The reacted proteins can then be detected by expression of the marker property on the resolved proteins of the electrophoresis gel.

A number of procedures along these general lines have been used to label or alter components situated on the outer aspect of the red cell membrane permeability barrier, and some of these are listed in Table 3.

D

(*N.B.* Many other group specific impermeable reagents have been used, e.g. p-chloromercuribenzene sulphonate, to combine with exposed —SH groups;[55] the listed compounds are those which have been used to identify proteins individually by electrophoresis.) When used to treat intact red blood cells the marker reagents label no more than two of the major membrane proteins; the internal material (haemoglobin) remains untouched. In contrast, when isolated membranes (which have no permeability barrier) are treated in the same way, then all of the membrane proteins are labelled. From this we see that all the red cell membrane proteins are exposed on one side of the membrane or the other: no protein is totally occluded within the phospholipid bilayer. Furthermore, the two proteins exposed on the outer surface of the membrane are also accessible at the interior surface; they actually span the membrane.[56, 57]

These conclusions rest entirely on the assumption that the attacking reagents are unable to enter the cell, either by permeation of the membrane, or by a pinocytotic mechanism. In the cases cited above, these assumptions would appear to be valid, but this need not be so in other cell types. Problems of permeation and pinocytosis can sometimes be overcome by attaching the reagent to a macroscopic carrier. Thus, p-mercuribenzoate, a permeant —SH reagent, was rendered impermeant but still active by coupling via the carboxyl group to chains of aminoethyl dextran having an average molecular weight of 150,000.[68] The coupled reagent was able to inhibit the ouabain-sensitive component of red cell membrane ATPase selectively and the site of attachment is probably related closely to the operation of the sodium pump. In order to obviate the possibility of internalisation by pinocytosis, enzymes can be attached to beads of modified agarose, having dimensions larger than the cells which they are to react with. When guinea-pig polymorphonuclear leukocytes (average diameter 10 μm) were treated with neuraminidase bound to agarose beads (40–120 μm) all the enzyme-reactive sialic acid was released. There was no possibility of internalisation of the neuraminidase, and it could be concluded that all the sialic acid is located at the membrane outer surface.[69]

PROTEINS WHICH SPAN THE RED CELL MEMBRANE[70]

(a) *Glycophorin*

By examining the labelling patterns of proteins derived from the membranes of labelled cells with those of isolated membranes, much can be learned about the disposition of proteins which span the membrane. Proteolytic digestion of the membrane surfaces is another useful approach. The major red cell glycoprotein (glycophorin, bands PAS 1 and PAS 2 on SDS polyacrylamide gels: see Fig. 19*b*) reacts with FMMP (Table 3) when intact red cells are treated with this reagent.[57] It is thus exposed at the membrane outer surface, but it reacts with far more of the label when the reagent is used to treat isolated membranes. Most importantly, it has been shown (by peptide analysis) that the label applied externally, or internally, attaches to distinctly different sites, and this clearly shows that the glycophorin is a transmembrane species, exposed at both surfaces of the membrane.

Glycophorin has been isolated.[71, 72] It is composed of only 87 amino acids and a very large proportion of these (over 25%) are the hydroxy amino acids serine and threonine. The remaining part (60%) of the molecular weight of 30,000 is made up of carbohydrate, of which there are more than 100 residues.

A number of peptides can be isolated from purified glycophorin by treatment with trypsin or by application of cyanogen bromide.[73] There is sufficient correspondence between the end sequences of these peptides to be able to place them in the order that they have in the native protein. These studies reveal that this protein is divided into three distinct functional and environmental zones which confer upon it the necessary amphipathic qualities which are needed to allow it to form stable structures with the distinctive sections of a simple phospholipid bilayer (see p. 82). The N-terminal region, which is exclusively labelled by marker reagents applied to the outside of the red cell permeability barrier, is the bearer of sugar and sialic acid residues. Only this section of the glycoprotein is attacked when intact red cells are incubated with trypsin. This portion carries the blood group antigens (ABO and MN blood group determinants) and lectin and influenza virus receptors. The carbohydrate residues are attached as a number of short oligosaccharide chains generally terminating in sialic acid groups which are the chief determinants of the negative surface charge quality of the cell.[74] The C-terminal region of the protein which contains no detectable carbohydrate can only be labelled when the permeability barrier of the membrane is broken, and this implies that this section of the protein is exposed on the intracellular aspect of the membrane. This region of the glycoprotein is characterised by having an unusually large proportion of proline residues.

From the central part of the protein, a water-insoluble peptide of 23 amino acid residues has been obtained. This peptide, which is not labelled by marker reagents applied to red cell membranes, contains less than 20% of amino acids of the type normally recognised as having charged side chains.[75] For these reasons it is concluded that the hydrophobic zone of the red cell glycophorin lies buried in the hydrocarbon central layer of the membrane. The only unfavourable contacts between a hydrophobic peptide and the hydrocarbon of phospholipid would be at the carbonyl and amino groups of the peptide linkage. However, these problems could be overcome if these groups form intramolecular hydrogen bonds which could stabilise the peptide in a helical conformation. It has been pointed out that as an α-helix, this section of glycophorin could form an "oily rod" of 35 Å in length, which would be sufficient to span the bilayer.[76] A diagrammatic representation of the probable organisation of glycophorin in the red cell membrane is illustrated in Fig. 20.

The glycophorin is probably anchored in the membrane to the intercalated particles which are revealed by freeze fracture electron microscopy.[77] We saw earlier (p. 77) how ferritin conjugates of antibodies to specific

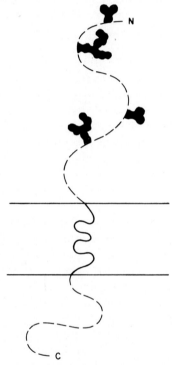

FIG. 20. Organisation of glycophorin in the red cell membrane

blood group substances were used to identify the intercalated particles as the sites of attachment of the blood group antigens, and now we have seen that isolated and purified glycophorin is itself the carrier of antigenic determinants. By treating red cell membranes with a ferritin-conjugated lectin (which attaches specifically to carbohydrate residues on the cell surface) and then arranging the conditions so as to aggregate the intercalated particles of the fracture plane before freezing and fracturing, it was possible to show that the labelled lectin binds to the outer etch face (PS) only in the roughened regions which override the aggregated particles of the fracture face (PF).

(b) The Band 3 Proteins
The other transmembrane component which has been identified by

labelling reagents and enzyme catalysed alteration accounts for about a quarter of the total red cell membrane protein.[70] It appears as rather a diffuse band on SDS-polyacylamide electrophoresis and it is possible that it is a heterogeneous mixture of proteins having molecular weights in the range 90,000–105,000. It is often called band 3, because of its position in SDS polyacrylamide gels[78] (sometimes component a). Although labelling experiments indicate that the band 3 protein is accessible on the exterior of the membrane, it is insensitive to tryptic digestion and is only partially hydrolysed following treatment of intact cells with pronase. Probably only a small proportion of the protein is exposed. Likewise, only a small number of peptides are accessible to labelling reagents directed at the internal surface. The main portion of this protein is shielded from attack by labelling reagents and trypsin applied from both the interior and the exterior aqueous phases, and for this reason it has been suggested that it lies largely buried inside the membrane.[70, 79] It is likely that this component has a compact globular structure, and on the basis of its probable dimensions and other indirect evidence, it has been suggested that it also shares identity with the intercalated particles of 80 Å diameter observed on the outwardly directed (PF) face in freeze fracture electron microscopy.

PROTEINS WITH DEFINED FUNCTION

Much can be learned, especially about protein-lipid interactions, from the study of isolated membrane proteins, but it is true to say that if transport and other controlled phenomena are to be understood at a molecular level, then the proteins themselves must be identified and isolated. Attempts have been made to tag individual proteins with labelled inhibitors of defined membrane functions, and by enzyme catalysed transfer of active groups to functional loci in the membrane. These approaches have been used in attempts to mark and identify intermediates of the sodium and potassium dependent ATPase complex of cell membranes.[80, 81] Another well-known example is the attempt to isolate post-synaptic acetylcholine receptor sites, by labelling the receptor with the potent antagonist ^{131}I-α-bungarotoxin.[45] The iodine label is then used as a marker in the later stages of membrane solubilisation and protein separation.

ISOLATION AND RECONSTITUTION OF THE ANION EXCHANGE PATHWAY OF THE HUMAN RED CELL MEMBRANE

A good example of the use of affinity labels is the recently reported partial purification and reconstitution of proteins involved in the exchange of anions (Cl^-, HCO_3^-, SO_4^{2-}, etc.) across the red cell membrane. The exchange of anions may be regarded as one of the main activities of this particular membrane as it enables the cell to dispose of CO_2 as bicarbonate (exchanging for chloride) as it moves from the tissues to the lungs. Chloride exchange in the red cell is so rapid ($t_{\frac{1}{2}} = 0.2$ sec at $23°$)[82] that it is normal to make measurements with less rapidly permeating inorganic ions such as SO_4 which are thought to utilise the same pathway.[83, 84] Anion exchange, a saturable one-for-one activity is quite distinct from the unidirectional flux of anions, which occurs exceedingly slowly in red blood cells.[85, 86] About the details of the structures controlling the process, little is known, but obviously, a clearer understanding might be possible if the material comprising the pathway could be obtained free of other membrane components, and then reconstituted in an active form by incorporation in a model membrane of controlled composition.

Some of the impermeant labelling reagents (see Table 3) have been found to have specific effects on membrane transport properties. Among amino group directed reagents (which include the well-known fluorodinitrobenzene and trinitrobenzene sulphonate which have long been used for

4,4'-diisothiocyano-2,2'-stilbene disulphonate DIDS(^{3}H)

amino end-group analysis of proteins and which have rather non-specific effects on membrane function), SITS is a specific inhibitor of the anion exchange pathway of the red cell membrane.[62] An analogue of this substance, DIDS(^{3}H) having two isothiocyanate groups, is a still more potent inhibitor and this has been used as an affinity label for the proteins of the anion exchange pathway.[87] As related above for other labels, treatment of membranes with DIDS(^{3}H) results in extensive labelling of proteins subsequently separated by polyacrylamide gel electrophoresis, but only a single "band" of protein reacts with the label when intact cells are used. The extent of labelling of this protein bears a linear relationship to the degree of inhibition of anion permeability, saturation of the cell with DIDS(^{3}H) (about 300,000 molecules per cell) resulting in almost complete inhibition.[88] The labelled protein (mol. wt. about 95,000 daltons) is a transmembrane protein as it takes up additional label when this is applied to isolated membranes. The protein, identified as involved in anion permeation on the basis of its reaction with DIDS(^{3}H) is also reactive in other procedures designed to mark exposed proteins on the surface of the cell such as lactoperoxidase catalysed iodination, pronase digestion, and labelling with other impermeant reagents. Very likely, the DIDS(^{3}H) reactive protein is identical with at least some component of the diffuse "band 3" identified as transmembrane material by labelling and poly-acrylamide gel electrophoresis. Labelling proceedures related to other membrane transport functions point to the involvement of the band 3 proteins in other activities. As an example of this it is possible to phosphorylate band 3 with ^{32}P-ATP in the presence of Mg^{2+} and Na^+, and this is a criterion for its involvement in the operation of the sodium pump.[80, 81, 89]

When red cell membranes are treated with the non-ionic detergent Triton X-100, considerably less protein is solubilised than following treatment with SDS[43, 44, 55]; the milder treatment is however sufficient to release the band 3 protein which is liberated preferentially (see Fig. 19 (c)). This fraction when incorporated into liposomes generates an anion pathway ($^{35}SO_4^{2-}$ flux is enhanced 3–10-fold) which can be inhibited by application of the inhibitor, DIDS(^{3}H). On the other hand, band 3 protein derived from cells which have been treated with the inhibitor before extraction with Triton X-100 are quite inactive as inducers of anion permeation in liposomes.[48] Clearly, a number of criteria have to be satisfied when attempting the isolation, purification and reconstitution of any membrane functional entity. By virtue of its activity and sensitivity to inhibition, the reconstituted anion exchange pathway which contains only phospholipid and the band 3 proteins would seem to meet these adequately.

MOBILITY OF SURFACE COMPONENTS[90]

An alternative way of identifying specific loci in cell membranes is by the use of labelled antibodies and plant lectins. As was indicated earlier (see the section on freeze-fracture electron microscopy) this approach allows us to investigate the disposition of membrane antigens in the plane of the membrane surface.

By the use of fluorescein[91] or ferritin[92] labelling, one may visualise ligands (immunoglobulin or lectin, e.g. concanavalin A) attached to the cell surface by fluorescence or electron microscopy. When a fluorescein-labelled antibody is attached to surface antigens of lymphocytes, the initial appearance of the green fluorescent material is diffuse and evenly spread around the periphery of the cell. This indicates that the surface antigens are randomly distributed. If observation is continued for a few minutes however, the fluorescent material first aggregates into a number of "patches" and eventually migrates to a unique pole of the cell to form a "cap". These phenomena can also be detected at a much higher degree of resolution by the use of ferritin-labelled ligands and electron microscopy (see Fig. 21). We considered earlier (page 28) much evidence suggesting fluidity in the plane of the phospholipid bilayer of model and biological membranes. The movement of labelled ligands following their attachment to the cell surface antigens is a clear indication of the mobility of membrane

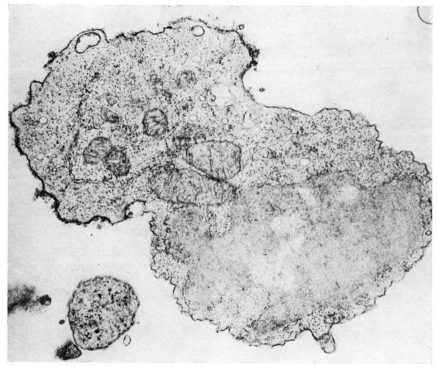

FIG. 21 (*a*). Extended incubation of lymphocytes with divalent antibody results in the generation of a cap over the end of the cell furthest removed from the nucleus.

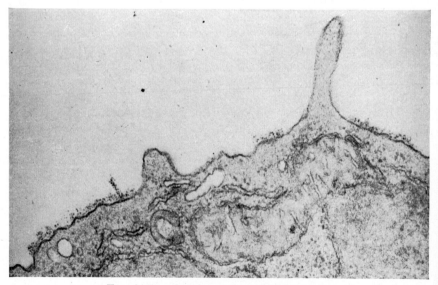

FIG. 21(*b*)—*Caption at foot of facing page.*

components in the two dimensions of this fluid plane. "Capping" and "patching" appear to be independent events. Capping (Fig. 21 (*a*)), which is not observed in all cell types, depends on prior patch formation, and is dependent on active metabolic processes within the cell. It may be prevented by the use of azide, (N_3^-) as a mitochondrial inhibitor. Patching (Fig. 21 (*b*)), on the other hand, is a passive process and depends on the valency of the immunological ligand used. The prime requirement here is for multivalency on the part of the antibody. The intact immunoglobulin molecule which has two Fab recognition subunits is divalent. It may attach to two separate antigenic structures on the cell surface to form $Ab.Ag_2$. In a like sense, proteins generally express antigenicity at more than one region of their polypeptide chains, and so it is possible to form massive aggregates of the type $Ab_m.Ag_n$. Patches may be regarded as surface precipitates; these depend for their formation on the ability of intact divalent antibodies to crosslink the antigenic structures of the cell surface. By treating immunoglobulin with the proteolytic enzyme papain, one may isolate single Fab fragments. These are still capable of attachment to antigens:

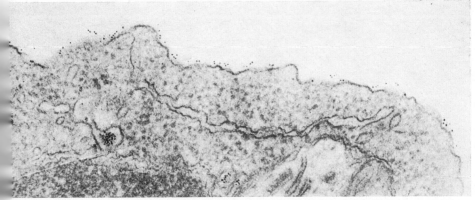

21 (*c*). Random distribution of receptors on a mouse B lymphocyte revealed by treatment ferritin labelled Fab (monovalent) fragments prepared from antibody directed to mouse ınoglobulin which is present on the B cell surface. There always remains the possibility ı small degree of redistribution may have occurred, as there is always a tendency for Fab ırations to dimerise and reform crosslinking units. Another complication is that the cell ce antigen (mouse immunoglobulin) is polyvalent, and may accommodate more than a single nolecule directed against different determinants in its structure. Nevertheless, the situation y different from that generated by treatment of B cells with intact antibody (Fig. 21 (*b*)). ı ref. 92.) (*d*) and (*c*) reproduced by permission of de Petris and Raff (1973), *Nature*, **241**,

FIG. 21 (*b*). Treatment of lymphocytes with divalent antibody, which crosslinks the surface Ig molecules, results in their redistribution into discrete patches after a few minutes (at 37°) leaving large areas of the cell surface free of the label. Patching is independent of the metabolic process of the cell and is merely slowed down, but not stopped by cooling. (From ref. 92.)

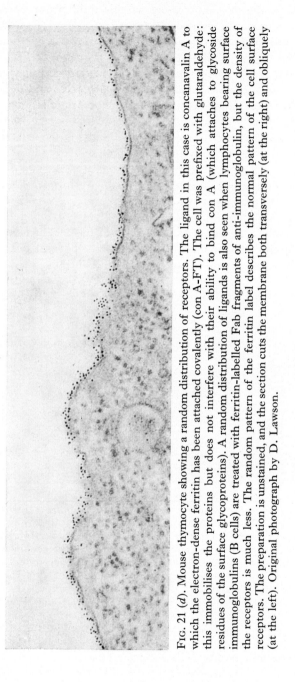

FIG. 21 (d). Mouse thymocyte showing a random distribution of receptors. The ligand in this case is concanavalin A to which the electron-dense ferritin has been attached covalently (con A-FT). The cell was prefixed with glutaraldehyde: this immobilises the proteins but does not interfere with their ability to bind con A (which attaches to glycoside residues of the surface glycoproteins). A random distribution of ligands is also seen when lymphocytes bearing surface immunoglobulins (B cells) are treated with ferritin-labelled Fab fragments of anti-immunoglobulin, but the density of the receptors is much less. The random pattern of the ferritin label describes the normal pattern of the cell surface receptors. The preparation is unstained, and the section cuts the membrane both transversely (at the right) and obliquely (at the left). Original photograph by D. Lawson.

Fab fragments are, however, single recognition units and are in consequence monovalent. They are incapable of crosslinking. Binding of Fab fragments to lymphocytes may be detected by the use of suitable labelling procedures (fluorescein or ferritin) and it is then found that the distribution of the monovalent antibody on the cell surface is always diffuse: there is no evidence for patching (see Fig. 21 (c) and (d)).

Apart from the demonstration of mobility of surface antigens on the lymphocyte membrane, the condition of multivalency required to generate patches of labelled Ag.Ab complexes is shared by a number of cellular activation processes, which are triggered by the attachment of specific "messengers" to receptors on the cell membrane. An example of a cellular activation process requiring a multivalent messenger will be discussed in the final chapter.

THE RED CELL MEMBRANE—A MODEL BIOLOGICAL CELL MEMBRANE?

It has been one of the aims of this account to demonstrate certain common features which underlie the structure and function of biological membranes, and of these, the concept of the phospholipid bilayer is outstanding. At no point in membrane science has evidence been found to shed real doubt upon the reality of the phospholipid bilayer as the central structural feature of biological membranes. Nonetheless, it must be said that the real background to the main concepts of membrane biology rests on the evidence of very few cell types, some of which lie at extremes in terms of composition and behaviour. In general, these have been selected for study because of the ease with which they can be manipulated: the choice of the membranes derived from red blood cells, myelin and *Acholeplasma laidlawii* illustrates how even when we turn from the model systems of Chapter 1, we seek the "model" membranes provided by Nature herself. In particular, the membrane of the non-nucleated mammalian red blood cell has often tended to become synonymous with every notion that we apply to our science. There is no justification for this uncritical approach, and there is some evidence that one should be as wary of extending the concepts derived from a study of the properties of the red cell membrane to the properties of other (nucleated) cell types, as one would naturally be of extending those derived from the models of Chapter 1.

Fluidity in the Plasma Membrane of Nucleated Cells
To illustrate the need for caution, one might want to consider why in the discussion of the patching-capping phenomenon, we turn away from the

red blood cell and introduce the lymphocyte membrane as an object for study. The patching of cell surface receptors by multivalent ligands such as plant lectins and appropriate immunoglobulins is a function not only of lymphocytes but also of such diverse types as mast cells[93] (see Chapter 4), basophilic leucocytes,[94] myelin[95] and synaptic membranes[95]. Patching, a consequence of crosslinking of membrane surface antigens by multivalent ligands, is accepted to be an indication of the rather unrestricted freedom which some membrane proteins enjoy, and which allows lateral movement in the plane of the membrane; but it is a phenomenon which has not been observed in the membrane of intact and fully viable red blood cells. Certainly, it is possible to aggregate the intercalated particles in the fracture plane of isolated red cell membranes, and when this occurs, certain determinants on the cell surface move in concert, as has been described (p. 78). In contrast, when lymphocytes are treated with antibody, so as to "cap" the surface antigens at one pole of the cell, there is no concomitant movement of the membrane intercalated particles, which remain randomly distributed in the fracture plane.[96] It would appear that the surface receptors of the nucleated cell are free to move, and that there is no stable link between them and the relatively immobile particles.

The most striking demonstration of protein mobility in the plane of a membrane concerned the mixing of surface antigens (subsequently detected individually by attachment of specific fluorescent antibodies) following the fusion of two cell types to form heterokaryons.[97] Within 40 min of fusing cultured cells of mouse and human origin by treatment with Sendai virus, almost total mixing of species specific surface antigens occurs. These could be visualised by application of fluorescein-labelled anti-mouse immunoglobulin (green fluorescence) and rhodamine labelled anti-human immunoglobulin (red fluorescence). The only way in which mixing of the surface antigens could be impeded was by lowering the temperature: metabolic blockade to inhibit respiration or protein synthesis was without effect. The movement of antigens on the heterokaryon, followed by labelling with fluorescent antibodies, provides us with a method of measuring the rate of diffusion of proteins in the membranes of different cells. By this method, a diffusion coefficient of $0\cdot2.10^{-9}$ cm^2 sec^{-1} has been estimated for the mouse-human heterokaryons, and a figure ten times higher has been obtained for cultured muscle cells.[98] Another approach has been to bleach intrinsic or extrinsic spectral markers on just a very small area of membrane surface, by very intense irradiation using a highly collimated beam in a microscope. The subsequent relaxation which leads to the re-establishment of an evenly coloured surface is then followed under attenuated illumination. The time taken to attain homogeneous colour density after localised bleaching of rhodopsin in the disc membranes of frog retinal rods is short,[99] with $t_{\frac{1}{2}}$ about 35 sec at 20°. In total contrast to this picture of rapid movement is the human red cell, where relaxation following local bleaching of covalently bound fluorescein was not detectable in a period of 20 min.[100] In this case, the fluorescein label was attached mainly to proteins of the band 3 complex, and so these results suggest that these proteins (which are associated with a number of transport functions) are held rigidly in fixed positions in the membrane.

The Red Cell Membrane—A Tough Viscoelastic Solid

By comparison with other cell membranes, that of the red cell appears to lie at an extreme of rigidity, and on the basis of mechanical studies has even warranted the description of a "tough viscoelastic solid".[101] Enough has been said in this and the previous chapter to exclude the possibility that this special property of the red cell membrane derives from differences in phospholipid bilayer composition or organisation. Whilst the structural basis for the rigid nature of the red cell membrane is not yet certain, it seems most probably to derive from the attachment of a fibrous protein with the name spectrin (spectrin is synonymous with tektin A,[102] which was formerly thought to be distinct). Spectrin, which comprises about 25% of the membrane protein, can be released from red cell membranes simply by reducing the ionic strength to very low levels or by the application of chelating agents.[103, 104] These are not the kind of treatments which would be expected to dislodge the intrinsic proteins which exist in close association with the hydrophobic inner structure of the membrane and so spectrin has been described as a peripheral, or extrinsic, membrane protein. It has been obtained in pure form[102] and found to be composed of a number of similar (but not identical) polypeptide chains linked together in a form which resists a variety of harsh treatments known to break down the secondary, tertiary and quarternary structural arrangements of most proteins.[102] On SDS-polyacrylamide electrophoresis of solubilised extracts of red cell membranes, it shows up as the two slowest moving components, bands 1 and 2, having molecular weights of 220,000 and 240,000. It has been shown, by the use of ferritin-labelled specific antibodies to spectrin, that its sites of attachment to the membrane are confined to the inner surface.[105] There is much evidence which suggests that spectrin exists in close association with the intercalated particles of the red cell membrane, and their associated proteins, such as glycophorin, and that the spectrin regulates their topographical distribution in the plane of the membrane.[105] Certainly, procedures which lead to the aggregation of purified spectrin (depressed pH; addition of antispectrin[106, 107]) also lead to clustering of intercalated particles and of the exposed sialic acid residues of glycophorin on the membrane outer surface. Conversely, procedures which tend to aggregate the surface sialic acid residues (e.g. application of a lectin derived from castor bean, *Ricinus communis* extracts) result in organisational perturbations which can be detected by an enhancement in the ability of a

bifunctional reagent (dimethyl malonimidate) to crosslink spectrin located
at the inner membrane surface.[108] Effects of aggregation at the cell outer
surface may be transmitted right through to the cell interior. The principles
underlying the manipulations which have been used to establish the
relationship between cell surface glycoproteins and the spectrin at the
inner face are illustrated in Fig. 22.[109]

As well as constraining the lateral movement of the intrinsic proteins of
the red cell membrane, the spectrin may also be the mechanical determinant

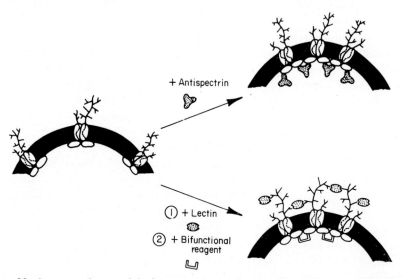

+ Antispectrin

① + Lectin
② + Bifunctional
    reagent

Fig. 22. A proposal to explain how treatment of red cell membranes with agents
which cause the aggregation of spectrin at the inner surface (addition of antispectrin antibody) results in the aggregation of glycophorin at the outer surface,
and how agents which crosslink glycophorin at the outer surface (lectins) lead to
the aggregation of proteins at the inner surface, as signified by an enhancement in
the ability of bifunctional ligands to crosslink spectrin. (From ref. 109.) Reproduced
by permission of Nicolson (1976), *BBA*, **457**, 57.

of the typical biconcave shape of some mammalian red blood cells. Its
inadvertent removal during the preparation of red cell membranes results in
some of the differences (such as mobility of intercalated particles) which
differentiate them from the original plasma membrane of the intact cell,
but it is possible to prepare membranes with the spectrin intact. The
relationship of a mechanical structural element with membrane proteins
is a specialisation of the red blood cell, and it may be this feature which
allows it to withstand the intense and prolonged buffeting and shearing
forces to which it is subject in the circulation.

The red cell, the membrane of which continues to be the focus of
almost overwhelming attention, is a primitive cell. It lacks a nucleus,

mitochondria, endoplasmic reticulum and all the functions associated with these structures. It also lacks storage granules of all kinds; there are no lysosomes, no secretory granules nor metabolic reserves (glycogen). The function of the red cell is unique: it is solely concerned with the transport of oxygen and carbon dioxide between the lungs and the tissues. The membrane of the cell which acts as a barrier to prevent the dissipation of the cell contents is endowed with functions primarily related to the upkeep of the interior milieu. Thus there exists a pathway for the entry of glucose, which is required for the generation of ATP and the reducing coenzymes glutathione and NADH which are involved in the maintenance of haemoglobin in the reduced (Fe II) form.

The ATP generated by glycolysis is largely consumed by the transport ATPases of the membrane which are responsible for the exclusion of Na^+ and Ca^{2+} from the cytosol, but in these terms the activity of the red cell membrane is slight in comparison with other cell types. The electrical conductivity[110] and the permeability of the red cell membrane to the small anions and cations[85, 86, 111, 112, 113] is low – not much greater than the corresponding parameters of the model membrane systems discussed in Chapter 1, and so the demand on the machinery of the transport ATPases is small. If the membrane of the red cell has a specialised function at all, then it is to be found in the pathway for anion exchange discussed earlier. The red cell membrane is evolved especially to enable bicarbonate and chloride ions to exchange with great rapidity, and this function is directly linked to the ability of the cell to transport CO_2.

The plasma membranes of most other cells have far more diverse functions. The ability to transport selected molecules is extended to the mechanism of endocytosis (i.e. pinocytosis and phagocytosis) whereby external material is first engulfed by the plasma membrane, and then included within a vesicle which can be carried into the interior of the cell. Another form of membrane transport allows for the release of intracellular material by the mechanism of exocytosis (or reversed pinocytosis): this is discussed further in Chapter 4. Localised developments of the plasma membrane associated with the presence of specialised structures allow for the recognition and attachment of individual cells to each other (synaptic and gap junctions) and so-called tight junctions (zonulae occludens) which allow for communication between cells by a low-resistance pathway. It is also the function of the cell membrane to receive and respond to chemical signals. These may be hormones, neurotransmitters or growth factors, or they may be exogeneous substances such as toxins, drugs or antigens. The membrane response to such signals may commit the cell to an inhibition or stimulation of metabolic processes, or such functions as contraction, endocytosis, exocytosis or proliferation. Despite the very great diversity of cellular activities, and the very large number of chemical messengers needed to enlist individual cells and tissues, it appears likely that there is only a very limited number of ways in which the receptors situated at the cell surface can act to transduce these external stimuli into signals to which the interior of the cell can respond. One such transducing mechanism forms the subject of discussion in the latter part of Chapter 4.

REFERENCES

1. Gorter, E. and Grendel, F. J. (1925). On bimolecular layers of lipoids on the chromocytes of the blood. *J. Exp. Med.* **41**, 439.
2. Mueller, P. and Rudin, D. O. (1969). Translocators in bimolecular lipid membranes. *In* "Current topics in Bioenergetics", Vol. 3, D. R. Sanadi (ed). Academic Press, London and New York.
3. Bar, R. S., Deamer, D. W. and Cornwell, D. G. (1966). Surface area of human erythrocyte lipids: reinvestigation of experiments on plasma membrane. *Science*, **153**, 1010.
4. Danielli, J. F. and Harvey, E. N. (1935). The tension at the surface of the mackerel egg oil with remarks on the nature of the cell surface. *J. Cell. Comp. Physiol.* **5**, 483.
5. Danielli, J. F. and Davson, H. (1935). A contribution to the theory of permeability of thin films. *J. Cell. Comp. Physiol.* **5**, 495.
6. *Davson, H. (1962). Growth of the concept of the paucimolecular membrane. *Circulation*, **26**, 1022.
7. Davson, H. and Danielli, J. F. (1943). "The Permeability of Natural Cell Membranes". Cambridge University Press.
8. *Robertson, J. D. (1959). The ultrastructure of cell membranes and their components. *In* "The Structure and Function of Subcellular Components", E. M. Crook (ed). Biochem. Soc. Symposium No. 16. Cambridge University Press.
9. Robertson, J. D. (1960). The molecular structure and contact relationship of cell membranes. *Prog. Biophys.* **10**, 343.
10. Robertson, J. D. (1967). The organization of cellular membranes. *In* "Molecular Organization and Biological Function", J. M. Allen (ed). Harper and Row, New York.
11. Robertson, J. D. (1964). Unit membranes. *In* "Cellular Membranes in Development", M. Locke (ed). Academic Press, London and New York.
12. Hendler, R. W. (1971). Biological membrane ultrastructure. *Physiol. Revs.* **51**, 66.
13. *Korn, E. D. (166). Structure of biological membranes. *Science*, **153**, 1491.
14. *Korn, E. D. (1968). Structure and function of the plasma membrane, a biochemical perspective. *J. Gen. Physiol.* **52**, 2575.
15. Fleischer, S., Fleischer, B. and Stoeckenius, W. (1967). Fine structure of lipid depleted mitochondria. *J. Cell. Biol.* **32**, 193.
16. Allt, G. (1975). The node of Ranvier in experimental neuritis: an electron microscope study. *J. Neurocytol.* **4**, 63.
17. Dowling, J. E. (1967). The organization of vertebrate visual receptors. *In*

---

* References marked with an asterisk (*) are mainly review articles especially recommended for further reading.

"Molecular Organization and Biological Function", J. M. Allen (ed). Harper and Row, New York.

18. Fernandez-Moran, H. and Finean, J. B. (1957). Electron microscope and low-angle X-ray diffraction studies of the nerve myelin sheath. *J. Biophys. Biochem. Cytol.* **3**, 725.

19. Finean, J. B. (1962). The nature and stability of the plasma membrane. *Circulation*, **26**, 1151.

20. Finean, J. B. (1966). The molecular organization of cell membranes. *Prog. Biophys. and Mol. Biol.* **16**, 143.

21. Caspar, D. L. D. and Kirschner, D. A. (1971). Myelin membrane structure at 10 Å resolution. *Nature New Biol.* **231**, 46.

22. *Chapman, D. and Dodd, G. H. (1971). Physiocochemical probes of membrane structure. *In* "Structure and Function of Biological Membranes", L. I. Rothfield (ed). Academic Press, London and New York.

23. Wilkins, M. F., Blaurock, A. E. and Engelman, D. M. (1971) Bilayer structure in membranes. *Nature New Biol.* **230**, 72.

24. Engelman, D. M. (1971). Lipid bilayer structure in the membrane of *Mycoplasma laidlawii*. *J. Molec. Biol.* **58**, 153.

25. Razin, S., Tourtellotte, M. E., McElhaney, R. N. and Pollack, J. D. (1966). Influence of lipid components of *Mycoplasma laidlawii* on osmotic fragility of cells. *J. Bact.* **91**, 609.

26. McElhaney, R. N. and Tourtellotte, M. E. (1969). Mycoplasma membrane lipids: variations in fatty acid composition. *Science*, **164**, 433.

27. *Branton, D. and Deamer, D. W. (1972). "Membrane Structure." Springer-Verlag, Vienna and New York.

28. Pinto da Silva, P. G. and Branton, D. (1970). Membrane splitting in freeze etching. *J. Cell. Biol.* **45**, 598.

29. Branton, D. and thirteen others (1975). Freeze-etching nomenclature. *Science*, **190**, 54.

30. Deamer, D. W. and Branton, D. (1967). Fracture planes in an ice-bilayer model membrane system. *Science*, **158**, 655.

31. Park, R. B. and Branton, D. (1966). Freeze-etching of chloroplasts from glutaraldehyde fixed leaves. *Brookhaven Symp. Biol.* **19**, 341.

32. Branton, D. and Park, R. B. (1967). Subunits in choroplast lamellae. *J. Ultrastruct. Res.* **19**, 283.

33. James, R. and Branton, D. (1971). The correlation between the saturation of membrane fatty acids and the presence of membrane fracture faces after osmium fixation. *Biochim. Biophys. Acta*, **233**, 504.

34. Wartiovaara, J. and Branton, D. (1970). Visualization of ribosomes by freeze-etching. *Exp. Cell. Res.* **61**, 403.

35. Branton, D. and Clark, A. W. (1968). Fracture faces in frozen outer segments from the guinea pig retina. *Z. Zellforsch.* **91**, 586.

36. Branton, D. (1967) Fracture faces of frozen myelin. *Exp. Cell Res.* **45**, 703.

37. Deamer, D. W., Leonard, R., Tardieu, A. and Branton, D. (1970). Lamellar and hexagonal lipid phases visualized by freeze etching. *Biochim. Biophys. Acta*, **219**, 47.

38. Pinto da Silva, P. G. (1972). Translational mobility of the membrane intercalated particles of human erythrocyte ghosts. *J. Cell. Biol.* **53**, 777.

39. Pinto da Silva, P. G., Douglas, S. D. and Branton, D. (1971). Localization of A antigen sites on human erythrocyte ghosts. *Nature*, **232**, 194.

40. Kawahara, K., Kirschner, A. G. and Tanford, C. (1965). Dissociation of human CO-hemoglobin by urea, guanidine hydrochloride and other reagents. *Biochemistry*, **4**, 1203.

41. Hoogeveen, J. Th., Juliano, R., Coleman, J. and Rothstein, S. (1970). Water soluble proteins of the human red cell membrane. *J. Membrane Biol.* **3**, 156.
42. Maddy, A. H. (1964). The solubilization of the protein of the ox erythrocyte ghost. *Biochim. Biophys. Acta*, **88**, 448.
43. Steck, T. L. and Yu, J. (1973). Selective solubilisation of proteins from red blood cell membranes by protein perturbants. *J. Supramolec. Structure*, **1**, 220.
44. Yu, J., Fischman, D. A. and Steck, T. L. (1973). Selective solubilisation of proteins and phospholipids from red blood cell membranes by nonionic detergents. *J. Supramolec. Structure*, **1**, 233.
45. Miledi, R. and Potter, L. T. (1971). Acetylcholine receptors in muscle fibres. *Nature*. **233**, 599.
46. Racker, E. (1972). Reconstitution of a calcium pump with phospholipids and a purified Ca^{2+}-dependent adenosine triphosphatase from sarcoplasmic reticulum. *J. Biol. Chem.* **247**, 8198.
47. Warren, G. B., Toon, P. A., Birdsall, N. J. M., Lee, A. G. and Metcalfe, J. C. (1974). Reconstitution of a calcium pump using defined membrane components. *Proc. Nat. Acad. Sci.* **71**, 622.
48. Rothstein, A., Cabantchik, Z. I., Balshin, M. and Juliano, R. (1975). Enhancement of anion permeability in lecithin vesicles by hydrophobic proteins extracted from red blood cell membranes. *Biochem. Biophys. Res. Commun.* **64**, 144.
49. *Chapman, D. and Wallach, D. F. H. (1968). Recent physical studies of phospholipids and natural membranes: optical rotatory dispersion studies. Chapter 2, p. 176, *in* "Biological Membranes, Physical Fact and Function", Vol. 1, D. Chapman (ed). Academic Press, London and New York.
50. Wallach, D. F. H. and Zahler, P. H. (1966). Protein conformations in cellular membranes. *Proc. Nat. Acad. Sci.* **56**, 1552.
51. Urry, D. W. (1972). Protein conformation in biomembranes: optical rotation and absorption of membrane suspensions. *Biochim. Biophys. Acta*, **265**, 115.
51a. Wallach, D. F. H., Low, D. A. and Bertland, A. V. (1973). Membrane optical activity: Some facts and fallacies. *Proc. Nat. Acad. Sci.* **70**, 3235.
51b. Gordon, A. S., Wallach, D. F. H. and Straus, J. H. (1969). The optical activity of plasma membranes and its modification by lysolecithin, phospholipase A and phospholipase C. *Biochim. Biophys. Acta*, **183**, 405.
52. Daemen, F. J. M. (1973). Vertebrate rod outer segment membranes. *Biochim. Biophys. Acta*, **300**, 255.
53. Oesterhelt, D. and Stoeckenius, W. (1971). Rhodopsin-like protein from the purple membrane of *Halobacterium halobium*. *Nature New Biol.* **233**, 149.
54. Henderson, R. and Unwin, P. N. T. (1975). Three-dimensional model of purple membrane obtained by electron microscopy. *Nature*, **257**, 28.
55. *Steck, T. L. (1974). The organization of proteins in the human red blood cell membrane. *J. Cell Biol.* **62**, 1.
55a. van Steeveninck, J., Weed, R. I. and Rothstein, A. (1965). Localization of erythrocyte membrane sulfhydryl groups essential for glucose transport. *J. Gen. Physiol.* **48**, 617.
56. Bretscher, M. S. (1971). Major human erythrocyte glycoprotein spans the cell membrane. *Nature New Biol.* **231**, 229.
57. Bretscher, M. S. (1971). Human erythrocyte membranes: specific labelling of surface proteins. *J. Mol. Biol.* **58**, 778.
58. Maddy, A. H. (1964). A fluorescent label for the outer components of the plasma membrane. *Biochim. Biophys. Acta*, **88**, 390.

59. Berg, H. C. (1969). Sulfanilic acid diazonium salt: a label for the outside of the human erythrocyte membrane. *Biochim. Biophys. Acta*, **183**, 65.

60. Bender, W. W., Garan, H. and Berg, H. C. (1971). Proteins of the human erythrocyte membrane as modified by pronase. *J. Mol. Biol.* **58**, 783.

61. Phillips, D. R. and Morrison, M. (1970). The arrangement of proteins in the human erythrocyte membrane. *Biochem. Biophys. Res. Commun.* **40**, 284.

62. Knauf, P. A. and Rothstein, A. (1971). Chemical modification of membranes: effects of sulfhydryl and amino reactive reagents on anion and cation permeability of the human red blood cell. *J. Gen. Physiol.* **58**, 190.

63. Phillips, D. R. and Morrison, M. (1971). Exterior proteins on the human erythrocyte membrane. *Biochem. Biophys. Res. Commun.* **45**, 1103.

64. Arrioti, J. J. and Garvin, J. E. (1972). Selective labeling of human erythrocyte membrane components with tritiated trinitrobenzenesulfonic acid and picryl chloride. *Biochem. Biophys. Res. Commun.* **49**, 205.

65. Passow, H. (1971). Effects of pronase on passive ion permeability of the human red blood cell. *J. Membrane Biol.* **6**, 233.

66. Staros, J. V. and Richards, F. M. (1974). Photochemical labeling of the surface proteins of human erythrocytes. *Biochemistry*, **13**, 2720.

67. Staros, J. V. Haley, B. E. and Richards, F. M. (1974). Human erythrocytes and resealed ghosts: a comparison of membrane topology. *J. Biol. Chem.* **249**, 5004.

68. Ohta, H., Matsumoto, J. Nagano, K., Fujita, M., and Nakao, M. (1971). The inhibition of Na, K activated adenosinetriphosphatase by a large molecule derivative of p-chloromercuribenzoic acid at the outer surface of the human red cell. *Biochem. Biophys. Res. Commun.* **42**, 1127.

69. *de Pierre, J. W. and Karnovsky, M. L. (1973). Plasma membranes of mammalian cells: a review of methods for their characterization and isolation. *J. Cell. Biol.* **56**, 275.

70. *Bretscher, M. S. (1973). Membrane structure; some general principles. *Science*, **181**, 622.

71. Kathan, R. H., Winzler, R. J. and Johnson, C. A. (1961). Preparation of an inhibitor of viral hemagglutination from human erythrocytes. *J. Exp. Med.* **113**, 37.

72. Kathan, R. H. and Winzler, R. J. (1963). Structure studies on the myxovirus hemagglutination inhibitor of human erythrocytes. *J. Biol. Chem.* **238**, 21.

73. Marchesi, V. T., Jackson, R. L., Segrest, J. P. and Kahane, I. (1973). Molecular features of the major glycoprotein of the human erythrocyte membrane. *Fed. Proc.* **32**, 1833.

74. Eylar, E. H., Madoff, M. A., Brody, O. V. and Oncly, J. L. (1962). The contribution of sialic acid to the surface charge of the erythrocyte. *J. Biol. Chem.* **237**, 1992.

75. Segrest, J. P., Jackson, R. L. and Marchesi, V. T. (1972). Red cell membrane glycoprotein: amino acid sequence of an intramembranous region. *Biochem. Biophys. Res. Commun.* **49**, 964.

76. Segrest, J. P., Kahane, I., Jackson, R. L. and Marchesi, V. T. (1973). Major glycoprotein of the human erythrocyte membrane: evidence for an amphipathic molecular structure. *Arch. Biochem. Biophys.* **155**, 167.

77. Pinto da Silva, P. and Nicolson, G. L. (1974). Freeze-etch localization of concanavalin A receptors to the membrane intercalated particles of human erythrocyte ghost membranes. *Biochim. Biophys. Acta*, **363**, 311.

78. Fairbanks, G., Steck, T. L. and Wallach, D. F. H. (1971). Electrophoretic analysis of the major polypeptides of the human erythrocyte membrane. *Biochemistry*, **10**, 2606.

79. Bretscher, M. S. (1971). A major protein which spans the human red cell membrane. *J. Mol. Biol.* **59**, 351.

80. Knauf, P. A., Proverbio, F. and Hoffman, J. F. (1974). Chemical characterization and pronase susceptibility of the Na : K pump-associated phosphoprotein of human red blood cells. *J. Gen. Physiol.* **63**, 305.

81. Knauf, P. A., Proverbio, F. and Rothstein, A. (1974). Electrophoretic separation of different phosphoproteins associated with Ca-ATPase and Na, K-ATPase in human red cell ghosts. *J. Gen. Physiol.* **63**, 324.

82. Tosteson, D. C., (1959). Halide transport in red blood cells. *Acta Physiol. Scand.* **46**, 19.

83. Passow, H. (1964). Ion and water permeability of the red blood cell. *In* "The Red Blood Cell", C. Bishop and D. Surgenor (eds). Academic Press, New York and London.

84. Deuticke, B. (1970). Anion permeability of the red blood cell. *Naturwissenschaften*, **4**, 172.

85. Dalmark, M. and Wieth, O. (1972). Temperature dependence of chloride, bromide, iodide, thiocyanate and salicylate transport in human red cells. *J. Physiol.* **224**, 583.

86. Harris, E. J. and Pressman, B. C., (1967). Obligate cation exchanges in red cells. *Nature*, **216**, 918.

87. Cabantchik, Z. I. and Rothstein, A. (1972). The nature of the membrane sites controlling anion permeability of human red blood cells as determined by studies with disulfonic stilbene derivatives. *J. Membrane Biol.* **10**, 311.

88. Cabantchik, Z. I. and Rothstein, A. (1974). Membrane proteins related to anion permeability of human red blood cells. *J. Membrane Biol.* **15**, 207.

89. Avruch, J. and Fairbanks, G. (1972). Demonstration of a phosphopeptide intermediate in the Mg^{++} dependent, Na^+- and K^+-stimulated adenosine triphosphatase reaction of the erythrocyte membrane. *Proc. Nat. Acad. Sci.* **69**, 1216.

90. *Bretscher, M. S. and Raff, M. C. (1975). Mammalian plasma membranes. *Nature*, **258**, 43.

91. Taylor, R. B., Duffus, W. P. H., Raff, M. C. and de Petris, S. (1971). Redistribution and pinocytosis of lymphocyte surface immunoglobulin molecules induced by anti-immunoglobulin antibody. *Nature New Biol.* **233**, 225.

92. de Petris, S. and Raff, M. C. (1973). Normal distribution, patching and capping of lymphocyte surface immunoglubulin studied by electron microscopy. *Nature New Biol.* **241**, 257.

93. Lawson, D., Raff, M. C., Fewtrell, C. M. S. and Gomperts, B. D. (1975). Anti-immunoglobulin-induced histamine secretion by rat peritoneal mast cells studied by immunoferritin electron microscopy. *J. Exp. Med.* **142**, 391.

94. Becker, K. E., Ishizaka, T., Metzger, H., Ishazaka, K. and Grimley, P. M. (1973). Surface IgE on human basophils during histamine release. *J. Exp. Med.* **138**, 394.

95. Matus, A., de Petris, S. and Raff, A. (1973). Mobility of concanavalin-A receptors in myelin and synaptic membranes. *Nature New Biol.* **244**, 278.

96. Karnovsky, M. J. and Unanue, E. R. (1973). Mapping and migration of lymphocyte surface macromolecules. *Fed. Proc.* **32**, 55.

97. Frye, L. D. and Edidin, M. (1970). The rapid intermixing of cell surface antigens after formation of mouse-human heterokaryons. *J. Cell. Sci.* **7**, 319.

98. Edidin, M. and Fambrough, D. (1973). Fluidity of the surface of cultured

muscle fibers. Rapid lateral diffusion of marked surface antigens. *J. Cell. Biol.* **57**, 27.

99.   Poo, M. and Cone, R. A. (1974). Lateral diffusion of rhodopsin in the photoreceptor membrane. *Nature*, **247**, 348.

100.   Peters, R., Peters, J., Tews, K. H. and Bahr, W. (1974). A micro-fluorimetric study of translational diffusion in erythrocyte membranes. *Biochim. Biophys. Acta*, **367**, 282.

101.   Rand, R. P. (1964). Mechanical properties of the red cell membrane: visco elastic breakdown of the membrane. *Biophys. J.* **4**, 303.

102.   Fuller, G. M., Boughter, J. M. and Morazzini, M. (1974). Evidence for multiple polypeptide chains in the membrane protein spectrin. *Biochemistry*, **13**, 3036.

103.   Marchesi, V. T. and Steers, E. (1968). Selective solubilization of a protein component of the red cell membrane. *Science*, **159**, 203.

104.   Marchesi, S. L., Steers, E., Marchesi, V. T. and Tillack, T. W. (1970). Physical and chemical properties of a protein isolated from red cell membranes. *Biochemistry*, **9**, 50.

105.   Nicolson, G. L., Marchesi, V. T. and Singer, S. J. (1971). The localization of spectrin on the inner surface of human red blood cell membranes by ferritin-conjugated antibodies. *J. Cell. Biol.* **51**, 265.

106.   Nicolson, G. L. and Painter, R. G. (1973). Anionic sites of human erythrocyte membranes: antispectrin-induced transmembrane aggregation of the binding sites for positively charged colloidal particles. *J. Cell. Biol.* **59**, 395.

107.   Elgsaeter, A. and Branton, D. (1974). Intramembrane particle aggregation in erythrocyte ghosts: the effects of protein removal. *J. Cell. Biol.* **63**, 1018.

108.   Ji, T. H. and Nicolson, G. L. (1974). Lectin binding and perturbation of the outer surface of the cell membrane induces a transmembrane organizational alteration at the inner surface. *Proc. Nat. Acad. Sci.* **61**, 2212.

109.   Nicolson, G. L. (1976). Transmembrane control of the receptors on normal and tumor cells: cytoplasmic influence over cell surface components. *Biochim. Biophys. Acta*, **457**, 57.

110.   Lassen, U. V. (1972). Membrane potential and membrane resistance of red cells. Page 291 *in* "Oxygen Affinity of Hemoglobin and Red Cell Acid Base Status". M. Rorth and P. Astrup (eds). Munksgaard, Copenhagen.

111.   Gunn, R. B., Dalmark, M., Tosteson, D. C. and Wieth, J. O. (1973). Characteristics of chloride transport in human red blood cells. *J. Gen. Physiol.* **61**, 185.

112.   Hunter, M. J. (1971). A quantitative estimate of the non-exchange-restricted chloride permeability of the human red cell. *J. Physiol.* **218**, 49P.

113.   Kaplan, J. and Passow, H. (1974). Effects of phloridzin on net chloride movements across the valinomycin-treated erthrocyte membrane. *J. Membrane Biol.* **19**, 179.

3 | Inducing Ion Permeability

THE MOBILE ION CARRIERS

The central topic of this chapter is the consideration of substances which are able to induce ionic permeability in the inert model membranes of Chapter 1, and to enhance the ionic permeability of biological cell membranes. Most of the permeability inducing substances derive from species of the genus *Streptomyces*, which have been cultured by the pharmaceutical industry from soil samples gathered world-wide, in the search for materials having antibiotic activity. In each of the main sections of this chapter we start by reviewing briefly some experiments in which biological and model membrane systems have been treated with various classes of these permeability inducing substances, and then proceed to interpretations of their effects based upon a knowledge of their chemical structure.

VALINOMYCIN AND NIGERICIN

Effects in Biological Membranes

(a) *Rat Liver Mitochondria*. Figure 1 shows the effect of two such antibiotics, valinomycin and nigericin, on the metabolism of normally respiring rat liver mitochondria.[1] The traces of proton and K^+ movements were recorded using a pH electrode and a K^+ sensitive glass electrode. When valinomycin is added to the suspension, there is a movement of potassium into the mitochondria (detected by the electrode as a loss of K^+ from the medium) and an efflux of protons (detected as a depression of pH). The light scattering of the mitochondrial suspension decreases due to osmotic swelling of the mitochondria. This arises from the exchange of protons (which are covalently linked to ionisable groups within the mitochondria, and so make no osmotic contribution) for K^+ ions which make a net osmotic contribution. When the second antibiotic, nigericin, is added to the mitochondria the situation is reversed. The mitochondria lose K$^+$ and take up H$^+$. The light scattering decreases as the mitochondria shrink. These changes are essentially restricted to movements of K$^+$ and Rb$^+$. If the mitochondria are suspended in a solution containing Na$^+$ salts instead of K$^+$ salts, then these antibiotic substances have little effect. In essence, the difference between the two antibiotics, both of which accelerate ion movements, is that whereas valinomycin is capable of promoting

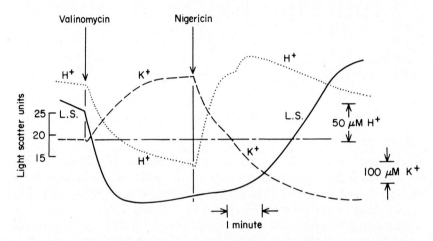

Fig. 1. Effects of valinomycin and nigericin on the transport of protons, potassium and water in respiring mitochondria. The movements of potassium (- - - -) and protons (.............) were followed with ion-specific glass electrodes, and the movement of water (———) was followed by measurements of light scattering. In the ion traces, a downward movement indicates an increase in the ion concentration measured by the electrode in question; i.e. a downward movement in the H$^+$ trace indicates a decrease in the pH of the medium and this is due to a loss of protons from the mitochondria. A downward shift in the light scattering trace indicates mitochondrial swelling due to an inward movement of water (see Chapter 1, p. 29). Reprinted from *Federation Proceedings*, **27**, 1283 (1968).

ion accumulation by respiring mitochondria, nigericin is only able to dissipate ionic gradients. If a respiratory inhibitor is added to the mitochondria, then valinomycin, too, acts to dissipate the gradient.

(b) *Red Blood Cells*. Human red blood cells which contain potassium as the principal cation leak only very slowly if deprived of substrate: they are very impermeable to ions and we might expect the K$^+$ loss to be accelerated if we add valinomycin, thereby introducing a pathway of potassium permeability. However, addition of valinomycin alone has little effect.[2] Only when valinomycin is added together with a weak organic acid which can carry protons across the membrane (such as dinitrophenol and other substances generally familiar as the uncouplers of mitochondrial oxidative

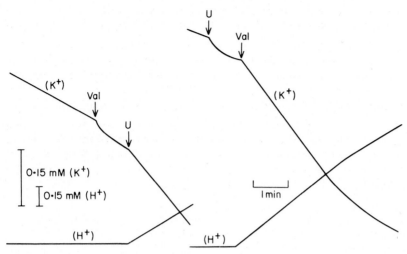

FIG. 2. The combined effects of valinomycin *plus* an uncoupler on the movement of K^+ out of red cells. The measurements were made with ion-specific glass electrodes. The red cells were suspended in a medium containing sucrose (0·3 M) with NaCl (0·01 M) and KCl (0·005 M). Notice that the addition of the individual compounds only produce a transient effect on the leakage of K^+ and have no effect on the movement of H^+. Addition of the two compounds is required to induce the exchange of internal K^+ for external H^+. Reproduced by permission of Harris and Pressman (1967), *Nature*, **216**, 918.

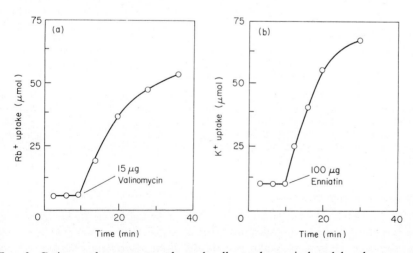

FIG. 3. Cation exchange across the red cell membrane induced by the neutral carrier antibiotics. (*a*) Valinomycin mediating exchange of internal K^+ for external $^{86}Rb^+$. (*b*) Enniatin mediating exchange of internal K^+ for external $^{42}K^+$. (From ref. 5.)

phosphorylation[3, 4] do we see a rapid outward movement of K⁺ with simultaneous inward movement of H⁺. Reversal of the order of addition, uncoupler first followed by valinomycin, also establishes the idea that in spite of providing a potential pathway for the movement of potassium, little net potassium movement actually occurs until a pathway for protons is provided too.[2]

Using isotopic tracers[5] (see Fig. 3), one may demonstrate exchange (without net flux) of K⁺ for *K⁺, K⁺ for *Rb⁺ or Rb⁺ for *Rb⁺ on addition of valinomycin, and this shows that the K⁺ leak pathway is not in itself dependent on the presence of the weak acid uncoupler substance.

In contrast, nigericin is sufficient by itself to induce net K⁺ efflux from red blood cells, in exchange for H⁺[2] (see Fig. 4).

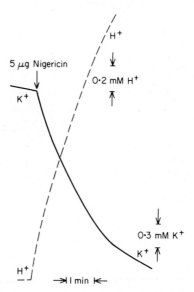

FIG. 4. Exchange of red cell K⁺ for external H⁺ induced by the anionic carrier antibiotic nigericin. Nigericin is effective without the addition of a subsidiary proton carrier (uncoupler); this figure should be compared with FIG. 2, p. 111. The measurements were made with ion-specific glass electrodes for K⁺ and H⁺. (From ref. 2.) Reproduced by permission of Harris and Pressman (1967), *Nature*, **216**, 918.

Effects in Model Membranes

(a) *Liposomes.* These substances act in a similar way to facilitate ion movements in model membranes.[5] In order to measure net ion fluxes in liposomes it is best to make use of the light-scattering changes which result from the shrinkage due to the exchange of internal cations for external protons (see p. 29. Chapter 1). Naturally, if an internal alkali cation is replaced by another alkali cation from the external solution, then

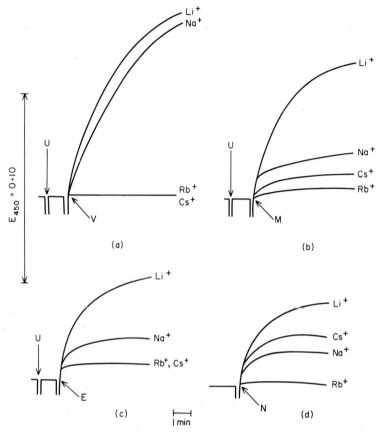

$E_{450} = 0 \cdot 10$

FIG. 5. The use of liposomes to test the selectivity of ion carrier substances. Liposomes were prepared from dried phospholipids (Chapter 1, p. 15) by hydration in KCl (0·04 M). They were then suspended in solutions of the other alkali cations at the same concentration, and monitored by light-scattering (optical density at an inert wavelength). In the absence of carrier antibiotics, the liposomes are exceedingly impermeable to any of the cations; there is no ion movement and consequently no movement of water and no change in the light-scattering properties of the suspension. An outward movement of K^+ from the liposomes was then induced by addition of the following agents:

(a) uncoupler (U) + valinomycin (V),
(b) uncoupler (U) + monactin (M), this is a close analogue of nonactin: see text,
(c) uncoupler (U) + enniatin (E), a depsipeptide resembling valinomycin,
(d) nigericin (N).

When the efflux of K^+ is exactly replaced by an influx of another cation, there is no movement of water and no change in light-scattering; when there is no counter-movement of alkali cations to balance the loss of K^+ (i.e. the loss of K^+ is balanced by an influx of H^+ which makes no osmotic contribution), then the liposomes shrink and the light-scattering increases, e.g. in FIG. 5 (a) the combination of uncoupler plus valinomycin which permits an efflux of K^+ does not permit a counter-movement of Li^+ or Na^+ and the liposomes shrink. Loss of K^+ is balanced by an influx of Rb^+ or Cs^+ so there is no net movement of H^+, no osmotic movement of water and no change in the light-scattering. Valinomycin can act as a carrier for K^+, Rb^+ and Cs^+, but not for Na^+ or Li^+.

there will be no shrinkage of the liposomes, and no change in the light-scattering properties of the liposomal suspension. The effect of valinomycin and nigericin on ion fluxes across liposome membranes is illustrated in Fig. 5. In this experiment, liposomes which had been loaded with a solution of KCl were suspended in equimolar solutions of the other alkali chlorides. Addition of nigericin is sufficient to cause an increase in the light-scattering (due to shrinkage) in the presence of Li^+, Cs^+ and Na^+. In the presence of Rb^+ there is no change and we can conclude that Rb^+ penetrates the nigericin treated liposomes as fast as K^+ leaks out. With the other ions we see graded responses which suggests that whilst the permeability of the treated membrane is highest for K^+ and Rb^+ there is a selectivity sequence with Na^+ preferred over Cs^+ and Li^+. Treatment of the KCl-loaded liposomes with valinomycin *alone* produces no effect on light-scattering. As with the previously described red blood cells, it is necessary also to add a reagent of the class used to uncouple mitochondrial oxidative phosphorylation, in order to facilitate the movement of protons across the membrane. When the membranes are treated with the two reagents together, then we see light-scattering changes due to the exchange of internal K^+ for H^+ in the presence of external Na^+ and Li^+ which are themselves unable to penetrate, but no light-scattering changes when Rb^+ and Cs^+ are the external cations: these are able to penetrate the valinomycin-treated membranes.

(b) *Phospholipid Bilayers.* The electrical properties of these permeability inducing substances may be studied by the use of phospholipid bilayer membranes. As was mentioned in Chapter 1, these are inert structures having extremely low electrical conductance and displaying no selectivity between the different alkali cations. The conductance of a phospholipid bilayer bathed by a solution containing K^+ or Rb^+ salts increases by about five orders of magnitude to around $10^{-4}\Omega^{-1}cm^{-2}$, following treatment with valinomycin.[6, 7] In the presence of the other alkali metal ions, the effect of valinomycin on conductance is slight. The ion-specific conductance induced by valinomycin shows strict adherance to Ohm's law ($V = IR$), the conductance ($1/R$) being directly proportional to the concentration of valinomycin. In contrast to the effects of valinomycin, treatment of bilayers with nigericin produces only a very slight increase in the electrical conductance,[6, 7] and this ion-specific permeability inducing substance is therefore said to be electrically silent.

If a solution of KCl (or RbCl) at two different concentrations is placed on either side of a bilayer membrane which is treated with valinomycin, then a potential will develop across the membrane. The magnitude of this potential (54–58 mV per decade concentration gradient) is very close to that predicted by the Nernst equation (for a derivation see ref. 8) for the current generated by a single ion species passing through a resistive membrane of high ionic selectivity, separating ionic solutions of different concentration

$$E = \frac{RT}{F} \ln \frac{C_1}{C_2},$$

where R = universal gas constant

T = absolute temperature

F = Faraday, the electric charge per gram equivalent of univalent ions

C_1 and C_2 are the concentrations of KCl on the two sides of the bilayer membrane.

At room temperature $(20°)$ $RT/F = 25$ mV, and if we convert to base-ten logarithms we get $E = 25.2 \cdot 3.\log C_1/C_2$. In an ideal situation with only one permeant ion at concentrations $0 \cdot 1$ M and $0 \cdot 01$ M.

$$E = 25.2 \cdot 3.\log 10$$
$$= 58 \text{ mV}$$

In the real situation, both the cation and the anion (K^+ and Cl^-) will tend to flow in a direction to dissipate the gradient. The sign of the potential which develops (in the case of the valinomycin-treated bilayer it is the high concentration side of the membrane which is positive) indicates that it is the positive ion (K^+, the cation) which is the primary carrier of current, and the close approach to ideality shows that the contribution of an anion current to the membrane potential is negligible. Once again, nigericin contrasts with valinomycin, and no potentials are developed in response to a concentration gradient of a transported ion.

If electrolyte solutions of different composition but similar ionic strength are placed on either side of a valinomycin-treated bilayer membrane, then the potential that develops will be the resultant of the potentials due to the two opposed ion flows. In this case we use the more general form of the Nernst equation:

$$E = RT \ln \frac{M_1^l + bM_2^l}{M_1^r + bM_2^r}$$

l and r refer to the left- and right-hand sides of the membrane, as in Fig. 6,

where the factor b expresses the relative permeability of ion M_2 compared with ion M_1 across the membrane. Obviously, if b is zero (membrane absolutely impermeable to ion M_2), then this expression reduces to the Nernst equation described above. This expression can be used to determine the relative permeability of the treated membrane for different ions. The membrane is made to separate two solutions composed of electrolytes M_1Cl and M_2Cl and arranged (as in Fig. 6) so that the total ionic concentration on either side of the membrane is the same, but so that there are opposing cationic gradients of equal magnitude across the membrane. In this example we have a solution of M_1Cl (0·1 M) and M_2Cl (0·01 M) on the left-hand side of the membrane, and a solution of M_1Cl (0·01 M) and M_2Cl (0·1 M) on the right-hand side of the membrane. If the membrane is now made permeable by treatment with one of the carrier substances, then there will be a tendency for the two separate ion gradients to dissipate, M_2^+

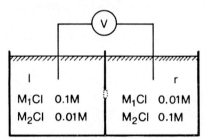

FIG. 6. Experimental arrangement for the measurement of ionic selectivity of current-carrying ionophores, using a lipid bilayer. For details, see text.

moving to the left hand-compartment and M_1^+ moving to the right-hand compartment. There is no gradient in the anionic component and so we do not have to take this into account in allocating the component parts of the membrane potential. Furthermore, by equating the ionic strength of the solutions we can reduce the differences which can arise from electrode potential effects.

Suppose in such an experiment, we find that $E = 40$ mV, with the left-hand side positive relative to the right. The direction of the potential informs us that the potential is primarily due to the passage of ion M_1 down its gradient (from left to right), and by making appropriate substitutions, we get

$$40 = \frac{RT}{F} \ln \frac{0·1 + 0·01b}{0·01 + 0·1b},$$

which gives $b = 0·1$; the mobility of ion M_2 through the membrane is one-tenth that of ion M_1.

For valinomycin-treated bilayers separating 0·05 M solutions of KCl and NaCl, resting potentials of up to 150 mV arise which are positive on the potassium side, showing that the potential results (mainly) from a flow of K^+ ions. The magnitude of the potential shows that the permeability ratio (K^+ over Na^+) is about 400.[6] By this kind of approach the ionic selectivity series $Rb^+ > K^+ > Cs^+ > Na^+ > Li^+$ for valinomycin-treated bilayers has been obtained. This method is much more precise than the light-scattering method using salt-loaded liposomes, but it is clear that the effect of valinomycin in enhancing ion fluxes in both systems is essentially similar.

Effects in Bulk Phases
Both valinomycin and nigericin are capable of extracting selected cations

into solution in organic solvents such as toluene, in which the solubility of the alkali cations is normally undetectable.[9] As we might now begin to predict from what has been said before, there is a difference in the manner whereby valinomycin and nigericin exert this effect.

In a standard experimental set-up, an aqueous solution of an alkali chloride together with a trace of the radioactive cation (generally it is preferred to work with $^{86}Rb^+$ than $^{40}K^+$ on account of a more suitable half-life of radioactive decay) is shaken together with an equal volume of organic solvent, such as toluene. The phases are separated, and a sample of the organic phase retained for counting.

The activity of the organic phase should be close to zero, indicating that no labelled cation has been extracted. If the experiment is now repeated in the presence of some nigericin, then it is possible to transfer radioactive material into the organic phase. In a similar manner to the red cell and liposome flux experiments, valinomycin is unable by itself to transfer cations into the bulk organic phase. However, if the chloride anion of the electrolyte is replaced by an organic-soluble anion, such as nitrophenolate, then one may demonstrate both the transfer of the radioactive cation, and a yellow colouration due to this anion, into the organic phase. It appears that the inability of valinomycin to facilitate the flux of cations across liposomes and red cell membranes resides in its requirement for an organic-soluble anion to render its complex with the carried cation soluble in the organic phase. With nigericin there is no such problem. We may conclude (and this will be demonstrated in greater detail later) that the complex of alkali cations with nigericin is neutral, whereas the complex of valinomycin with an alkali cation carries a positive charge, which must be neutralised if the complex is to enter the organic phase.

Extraction experiments of the type described above provide a simple way of determining the affinity of an ion carrier substance for different ions.[1] The limiting amount of a cation which can be extracted into an organic phase is of course limited by the amount of the carrier molecule present (e.g. valinomycin or nigericin). The affinity of the carrier for a cation is expressed by the concentration of the cation (in the aqueous phase) which achieves half-maximal transfer into the organic phase. This is most simply determined by the binding equation, most familiar in the form

$$[M^+]_{org} = \frac{[carrier]}{1 + \dfrac{1}{K[M^+]_{aq}}}$$

in which it resembles the Michaelis-Menten equation of elementary enzyme kinetics. By taking reciprocals we get

$$\frac{1}{[M^+]_{org}} = \frac{1}{[carrier]} + \frac{1}{K[M^+]_{aq}}$$

In the experiment we determine the amount of cation in both the aqueous and the organic phases over a range of different total cation concentrations. We plot the data in the form of a double reciprocal plot, which should give a straight line, having a y-intercept equal to the reciprocal concentration of the carrier substance used (i.e. the concentration of cation in the organic phase when the concentration in the aqueous phase is extrapolated to infinity) and which indicates 1:1 stoichiometry in the complexation reaction between the carrier and the cation. The negative

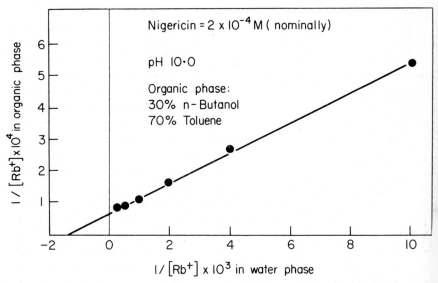

FIG. 7 (*a*). Extraction of ion-carrier complexes into organic solvents. A toluene-based organic phase containing nigericin was shaken with solutions containing $(Rb)_2SO_4$ at various concentrations and a trace of ^{86}Rb. After allowing the organic and aqueous phases to separate, the concentration of Rb^+ in each was estimated by liquid scintillation counting. The data are expressed as a double reciprocal plot. The stoichiometry of the K^+-Nig^- complex is about 1:1. (From ref. 1.)

intercept on the x-axis is equal in magnitude to the association constant, in units of reciprocal concentration. In practice, the relative selectivity of an ion carrier substance for different cations is generally determined by a competition experiment. Here the ability of a selection of different ions to transfer a single radioactive cation ($^{86}Rb^+$) from its carrier complex in the organic phase into aqueous solution is measured. Obviously, the cation which is active at the lowest concentration (generally this means displacing half of the stoichiometric amount of the labelled cation) is the preferred cation of the series. It was in this way that the selectivity series among alkali metal cations for various ion carrier substances (illustrated in Fig. 8) were determined. Generally the selectivity and affinity of ion carrier substances for the alkali metal cations is closely similar to the sequences of selectivity generated by the measurement of induced ion fluxes in liposomes and resting potentials in bilayers. Thus, the selectivity series for valino-mycin complexation and extraction into an organic phase consisting of

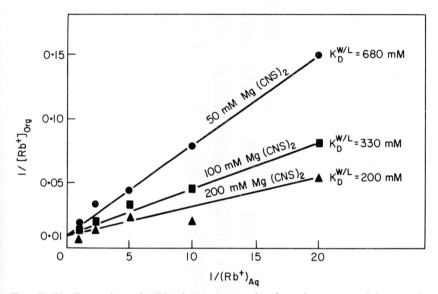

FIG. 7 (*b*). Extraction of [86]Rb$^+$ into an organic phase by a neutral ion carrier substance: the extraction is enhanced by the presence of a lipid soluble anion (CNS$^-$) but the stoichiometry of complex formation is constant, as shown by the common (extrapolated) intercept at "infinite" Rb$^+$ concentration. Reprinted from *Federation Proceedings*, **27**, 1283 (1968).

toluene (70%) and butanol (30%) is Rb$^+$ > K$^+$ > Cs$^+$ > Na$^+$ > Li$^+$, which is the same as that discussed earlier for bilayers. The similarity of the equilibrium (bulk phase determination) and kinetic (flux and potential measurements) properties suggests that the main kinetic barrier to induced ion transport is at the points of ion complexation and decomplexation rather than a restriction on the diffusion of the loaded carrier across the organic phase of the membrane.

The Ionophores: Structure and Function

We shall now examine the structures of valinomycin, nigericin and some related materials, in order to try and understand their permeability in-ducing capability. We should try and understand the ion specificity displayed, and also contrast the distinctive features of valinomycin, which mediates both conservative and dissipative ion movements with those of nigericin, which is purely dissipative in function. Substances inducing ion

E

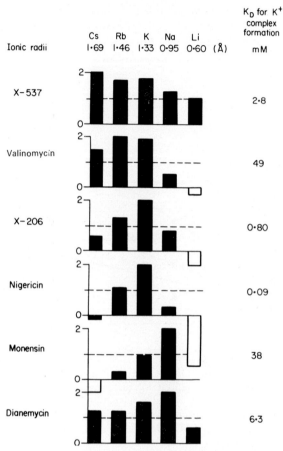

FIG. 8. Ion selectivity patterns for valinomycin and various anionic (nigericin-like) ion carrier substances. The data were mainly derived from the ability of the ions to displace ⁸⁶Rb⁺ from complexes with the antibiotics in two phase extraction experiments. The relative affinities are plotted on a log scale, so that a difference of 1 indicates a 10-fold preference of the carrier substance of one ion over another. It can be seen that nigericin has a one hundred fold preference for K⁺ over Na⁺, and that monensin has a tenfold preference for Na⁺ over K⁺. (From ref. 1.)

permeability of the types exemplified by valinomycin and nigericin, which exert their effects by acting as mobile ion carriers, are collectively termed ionophores (from φόρον bearer).[9] This term is also sometimes extended loosely to embrace all substances inducing ion permeability by whatever means.

Neutral Carrier Substances

Valinomycin is a cyclic depsipeptide (Fig. 9): that is to say a half peptide, containing equal proportions of peptide and ester linkages.[10] It is composed of four groups: two amino acids (D- and L-valine) and two hydroxy-acids (D-hydroxy isovaleric and L-lactic). The sequence of alternating amino and hydroxy acids is repeated three times to complete the ring. The side chains (methyl and isopropyl) are of low chemical reactivity. All

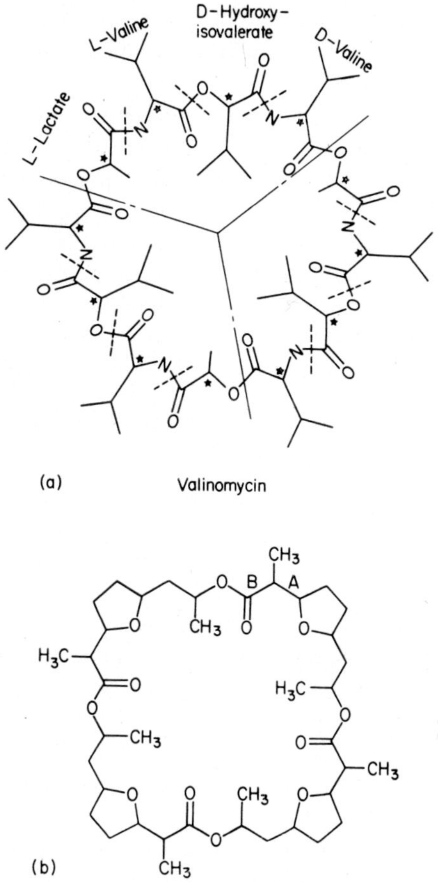

(a) Valinomycin

(b)

FIG. 9. (a) The primary structure of valinomycin. The asterisks indicate the optically active carbon atoms of the hydroxy and amino acids. (b) The structure of nonactin, another neutral ring, which shares many of the properties of valinomycin.

molecules of the valinomycin class are neutral rings and much evidence suggests that they are true ionophores, acting as mobile ion carriers. The carrier forms a complex with the carried ion at the aqueous interface of the membrane. It then diffuses across the hydrophobic central zone of the membrane to the aqueous interface at the distal side of the membrane, where the ion is released. In support of this principle we should consider some of the essential criteria for substances having the properties of an ionophore. Firstly, an ionophore should have the ability to combine with, and display selectivity among ions. Secondly, the complex of carrier and carried ion should be soluble in hydrocarbon solvents. Thirdly, the stability of the ion-carrier complex should not be so high that the carried ion is not readily released into aqueous solution at the distal side of the membrane. The kinetics of ion complexation and dissociation should be unhindered and rapid. These criteria are all satisfied by the ionophoric substances of the valinomycin class.

Structure of Nonactin. The first ionophore-metal complex structure to be solved at a crystallographic level was that of nonactin.[11] Indeed, so feature-less is this substance in terms of active chemical groups that X-ray diffraction proved to be the means of solving the primary structure too: this has been a common experience with many of the ionophores. In terms of its capacity to carry ions across lipid barriers, nonactin resembles valinomycin very closely; its structure is however much simpler. The presence of peptide linkages in valinomycin permits the formation of intramolecular hydrogen bonds, and depending on the polarity of the environment, different hydrogen-bonded structures (particularly of the uncomplexed molecule) can form. Because of this possibility, there is no *a priori* reason why the conformation of valinomycin in the crystal, and in solution within the phospholipid bilayer, should be the same. Indeed, there is spectral (NMR) evidence which suggests that the hydrogen-bonding forms of valinomycin in organic solvents are in fact different from the crystal form.[12] Because of the complexity and relative uncertainty of the conformation of valinomycin, consideration of three-dimensional structures will be restricted to the simpler nonactin, an entirely relevant analogue.

Nonactin possesses a 32-membered ring consisting of four repeating units joined by ester linkages.[11] The three-dimensional structure of the K^+-nonactin complex is illustrated in Fig. 10.[11] The potassium ion has been stripped of its hydration shell, and a nearly cubic coordination shell is now supplied by eight of the oxygen atoms of the ionophore molecule (4 carbonyl, 4 furan). In the complex, the distance between the atomic centres $K^+ \ldots O-$ is just the sum of the crystal radii, 2·7 Å, ensuring a very snug fit for the preferred ion. Due to the ion-induced dipole interactions between the cation and the eight oxygens, the charge on the ion is extensively delocalised in the molecule, and it is this feature, characteristic of the ionophores, which enables them to behave as alkali metal cation complexing substances. Lipid solubility is conferred on the complex by the chemical nature of the exterior surface which is composed entirely of lipophilic groups. This surface fully shields the polar interior cavity from the external environment. In the complexed molecule, the backbone chain is in the form of the seam of a tennis ball, to give the molecule a compact and approximately spherical shape ($15 \times 17 \times 12$ Å).

The structure of uncomplexed nonactin is illustrated in Fig. 11.[13]

It is less compact than the K⁺ complex ($17 \times 17 \times 8.5$ Å) and has a large hole in the centre: this allows the hydrated potassium ion to approach very close to the bonding oxygen atoms of the ionophore. As the potassium

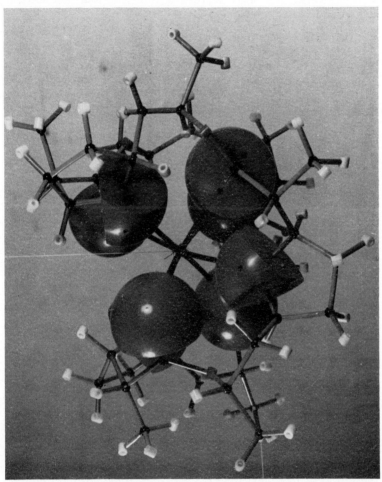

FIG. 10. Model of the (K̄-nonactin)⁺ complex. Only the bonding oxygen atoms are represented by space-filling shells so that the central cavity which is exactly filled by the potassium (black rods) can be seen.

comes into contact with the nonactin the water molecules of its hydration shell are rapidly shed one by one. The loss of hydration energy (about 80 Kcal mole⁻¹) is so great the the complexation process is only made thermodynamically and kinetically feasible by coupling the successive removal of single water molecules to compensatory conformational alterations of the ligand. This ensures that the coordination number of the ion is kept essentially constant as the aqueous hydration shell is replaced by

the coordination cavity of the nonactin[14] and is achieved by tortional rota-
tions at only four points (indicated as A and B in Figs 9 (*b*) and 11) in the
macrocyclic ring: no bonds have to be broken and no bonds have to be
made. The energy of activation for the process is low and there are no
major kinetic hurdles to be overcome.

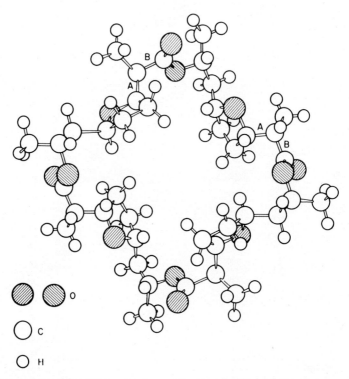

FIG. 11. The structure of nonactin as revealed by X-ray crystallography. This is an
open ring, which converts to the more compact formation of the K$^+$ complex by
torsional changes only at the four angles of the type marked A and B (see also
Fig. 9 (b)). The twelve oxygen atoms are in-filled with oblique lines: only eight
of these (oblique lines sloping up to the right) are involved in cation complexation.
Reproduced by permission of Dobler (1973), *Helv. Chem. Acta*, **55**, 1371.

All the neutral carrier substances are fully enclosed rings having no net
charge. The complexes therefore bear the single positive charge of the
cation. We can understand the behaviour and some of the restraints on these
substances on this basis. Thus, it is because the complex is positively
charged that it is necessary to provide a lipid-soluble anion in order for
the carrier to extract cations into organic solvents. Electroneutrality is
maintained in both the aqueous and organic phases by extracting equivalent amounts of the positively charged and lipid soluble complex and
the lipid soluble anion. Similarly, the electroneutrality principle applies

to the red cell which can only lose potassium in an electrically "silent" manner, by the exchange of K^+ for H^+, or by the simultaneous loss of K^+ and Cl^-. (*NB*. The permeability of the red cell membrane towards chloride ion can be enhanced by treatment with phlorizin[15]; in this

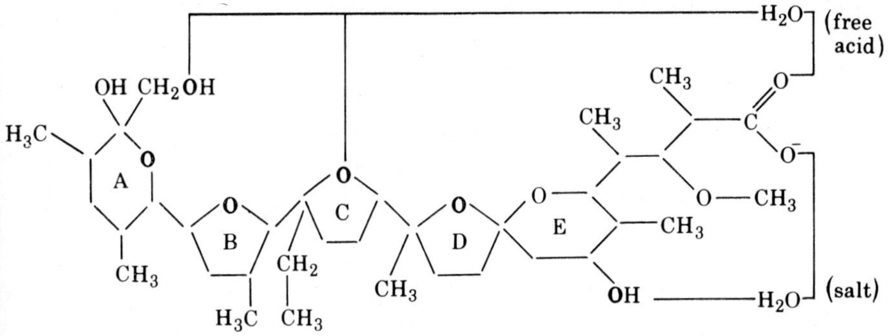

monensin

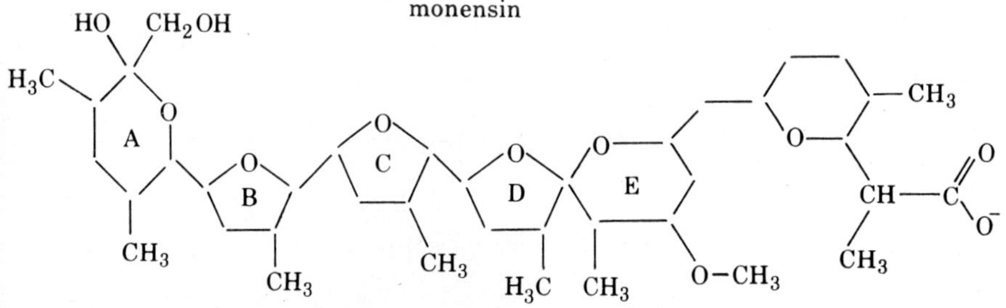

nigericin

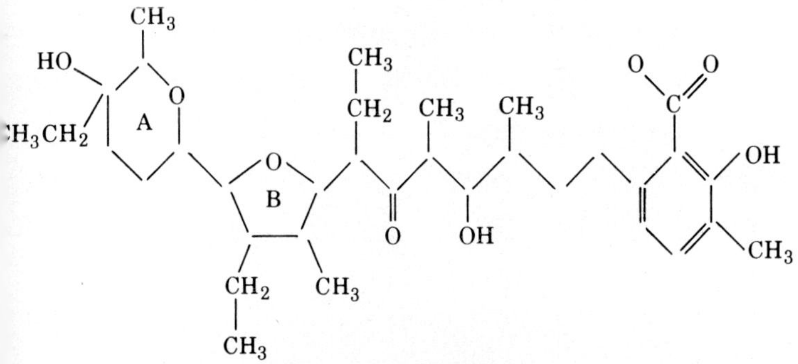

X537A

FIG. 12. The primary structures of the anionic ion carrier substances, monensin, nigericin and X537A. These all take up a cyclic configuration by hydrogen bond formation between the carboxyl group and hydroxyl groups on ring A. The three-dimensional structure is further stabilised (in the crystal) by hydrogen-bonded bridges which can involve water molecules. In the structure of monensin, the hydrogen bonded water bridges are indicated for the complex and for the free acid. The six bonding oxygens are printed in heavy type.

condition, application of valinomycin facilitates the loss of KCl, with consequent cell shrinkage.) In the case of two carried ions no such constraints apply, and this is why valinomycin and nonactin can effectively exchange K^+ for $*K^+$, K^+ for Rb^+, etc.

The increase in the conductance of lipid bilayers treated with neutral carriers in the presence of suitable cations results from the current carried down the applied potential gradient by the loaded complex. The unloaded neutral carrier diffuses back freely across the bilayer to gather up more cation, but carries zero current in the unloaded state. The bi-ionic potential set up across a bilayer having different ions on either side, or a common ion with a concentration difference, results from the sum of the ion currents in opposing directions.

Anionic Carrier Substances

We have seen that carriers of the nigericin class do none of these things. They are electrically insensitive and only act to dissipate ion concentration gradients, fulfilling the description of exchange diffusion carriers. The reason is not hard to see. The formulae of nigericin [16] together with monensin[17, 18, 19] and X537A[20, 21, 22] are illustrated in Fig. 12. These all

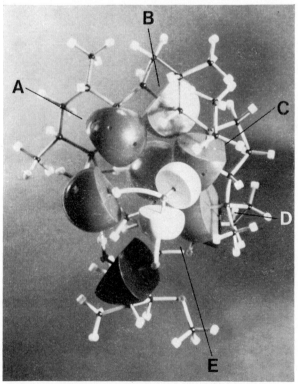

FIG 13 (a) Caption and FIG. 13 (b) on facing page.

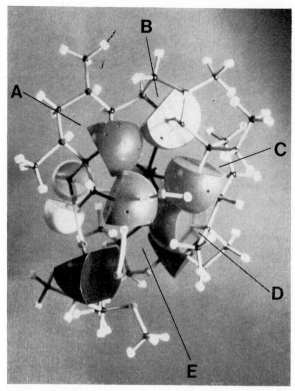

FIG. 13. Molecular models of monensin (*a*) and its Ag⁺ complex (*b*). (*N.B.* The Ag⁺ complex is isomorphous with the Na⁺ complex, but it is more amenable to crystallographic X-ray analysis.) The oxygen atoms involved in bonding with the cation, and also the carboxyl carbon (visible at the lower left in each photograph) and the "spiro" carbon atom which is shared by rings D and E, have been represented by space-filling shells to reveal the dimensions and shape of the central ion cavity. The hydrogen bonds joining the carboxyl group to ring A and to the bridging water molecule are represented by white bars. In the crystal of the Ag⁺ complex there are two molecules of water. One of these is merely appended to the bonding oxygen atom of the methoxy group of ring A and is probably an artefact of crystallisation; the other forms a hydrogen-bonded bridge across the open side of the molecule from the non-bonding hydroxyl of ring A to the bonding hydroxyl of ring E, and may play a role in the stabilisation of the complex configuration. The overall shape of the free monensin is very similar to the shape of the complex with even the central cavity being maintained: but this is no longer closed off by the water molecule, which instead bridges between the bonding oxygen of the methoxy group (ring A) and the bonding furan oxygen of ring C and the carboxyl group. There is an additional hydrogen bond connecting the bonding furan oxygen of ring D and the bonding hydroxyl oxygen of ring E which are in very close apposition. The arrangement of hydrogen bonds and bridging groups possibly holds the cavity of the free acid in a receptive conformation for an incoming cation.

have a single carboxylate group and so are negatively charged at neutral pH. The crystal structure of monensin and its silver salt (which is isomorphous with the Na^+ salt) is illustrated in Fig. 13 (*a*) and (*b*). In spite of a large number of oxygen atoms, all these compounds are extremely insoluble in water, and this must be due to the screening of the oxygens from the external environment which arises from the ring structure, formed by hydrogen bonding between the carboxylate group and the hydroxyl groups attached to the pyrian ring A. As with the neutral carriers, the cation is well shielded from the external environment, though in this case the geometry of the 6-fold coordination shell is irregular. The important functional difference is that the complexes of monensin, nigericin and X537A are neutral and therefore electrically insensitive. For this reason these substances are able to extract cations directly into lipid solvents without the additional requirement of a lipid soluble anion. We saw that mitochondria are unable to accumulate K^+ when treated with nigericin: this is because the loaded carrier, K^+-nig^- is neutral, and can only respond to concentration gradients. Nigericin alone is sufficient to release K^+ from red blood cells and liposomes. It does this by exchanging interior K^+ for exterior H^+ ions, and the exterior goes alkaline. The unloaded carrier remains neutral in the return journey through the membrane by associating with a proton. Monensin shows some degree of preference for Na^+. As with the previously discussed neutral carriers, the conformational rearrangements of monensin and nigericin during complexation with the carried cation have low energies of activation.

Carriers for Divalent Cations

Unlike the other ion carrier substances discussed so far, X537A is characterised by having a very broad range of cationic specificity.[23] Not only does it act as a carrier for all the alkali metal cations (see Fig. 8, p. 120) but it is also able to carry some divalent cations and primary amines (such as ethanolamine, noradrenaline and dopamine) into organic phases and across membranes. If we compare the structure of X537A with the more specific ion carriers nigericin and monensin, then we can see that X537A must have a much greater degree of conformational flexibility than its congeners. Rigidity in the chain of monensin and nigericin (and nonactin) is ensured by a succession of rings. This is particularly true in the region of rings D and E which are linked by a common carbon atom having the inflexible spiroketal configuration. It is probably these features which, by limiting flexibility and so setting conformational limits, ensure the high specificity towards individual cations of nigericin and monensin, and the broad range specificity of X537A. The conformational flexibility of X537A can be appreciated by considering the crystal structure of its complex with barium[20]. The divalent ion is held in asymmetrical nine-fold coordination by two molecules of the carrier substance and one molecule of water,

$Ba^{2+}(X537A^-)_2.H_2O$ and the complex as a whole is neutral in common with the other cation complexes of carboxylate ionophores. One X537A ligand offers six oxygen atoms to the divalent cation in a manner having some resemblance to the normal monovalent cation complexes of the carboxylate ionophores. The central cation is not totally enveloped, however, and is

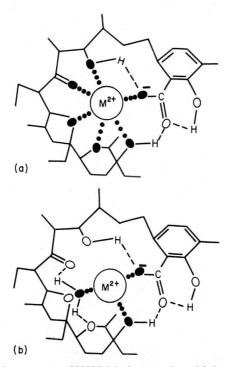

(a)

(b)

FIG. 14. Formalised structures of X537A in its complex with barium. One molecule offers six oxygen atoms to the complexed ion and is similar to the form of the ionophore ligands in their complexes alkali cations. The other molecule offers only two oxygens, plus the oxygen of a hydrogen bonded water molecule which bridges the ligand internally. There are no bridging structures between the two X537A molecules in the complex which are associated only through their interaction with the central divalent metal cation. Reprinted from *Federation Proceedings*, **32**, 1698 (1973).

accessible to a second ligand molecule, which presents just two oxygens and the hydrogen-bonded water, by taking up an entirely different configuration. There is no interaction by hydrogen bonds or any other bridging groups between the two X537A molecules, which are therefore held together solely by their interaction with the central cation. The formal configurations of X537A in its complex with Ba^{2+} are illustrated in Fig. 14.

Another linear carboxylate ion carrier substance which is able to transfer divalent cations through model and biological membranes and into bulk organic phases is A23187.[24] Unlike X537A, this substance shows little tendency to transport monovalent cations and amines and shows some selectivity between divalent cations. Among divalent cations transported by A23187 may be included the alkaline earth cations and Fe^{2+}.[25] The chemistry of A23187 which has pyrrole and benzoxazole ring structures

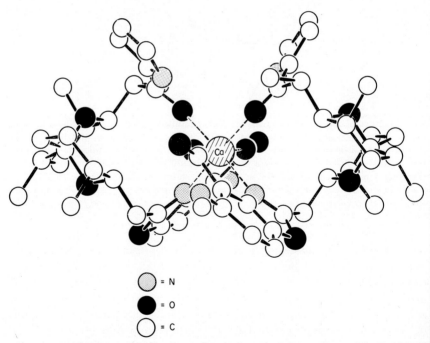

FIG. 15 (*a*). The primary structure of ionophore A23187. This molecule cyclises by hydrogen bond formation between the carboxyl and pyrrole groups. (From ref. 26.)

FIG. 15 (*b*). Three-dimensional crystal structure of the calcium salt of A23187. (From ref. 26a.) Reproduced by permission of Chaney (1976), *J. Antibiotics*, **29**, 4.

appears at first sight to be very different from all the other ionophores so far discussed,[26] but in the central region of the chain there are two pyran rings which are joined by the spiro linkage which is familiar from the other carboxylate carriers, and it is possible that the selectivity towards individual cations which this substance displays, derives (at least in part) from the conformational limitations imposed by this rigid structure. Metal ion complexes are of the type $M^{2+}(A23187^-)_2$. The crystal structure of the calcium salt of A23187 is illustrated in Fig. 15 (b)[26a]: it differs from the salts of X537A in a number of important respects. The calcium ion is held in regular octahedral coordination, both ligands bonding through the carbonyl group, the nitrogen atom of the benzoxazole ring, and the carboxyl group. The two ligands are further interconnected by direct hydrogen bonding from the pyrrole nitrogen atom of one molecule to the carboxyl oxygen of the other. The spiro-linked pyran rings are held well away from the central metal atom and play no direct part in the bonding of the central metal ion (cf. monensin, p. 127); their role appears to be concerned in the very precise positioning of the bonding atoms.

A23187 has been widely employed as a probe of calcium function in cellular activation phenomena, a topic which is considered further in Chapter 4.

PORES AND CHANNELS

So far we have considered ion transfer across hydrocarbon barriers solely in terms of carrier mechanisms. The ionophores are themselves mobile. Another way of inducing ion permeability is to place fixed pathways into the membrane, through which the ions can move.[27]

There is a wide range of substances, generally of microbiological origin, which may be used to induce permeability in this way. The chief feature is that they do not enclose individual ions in a lipid soluble parcel to be diffused across the membrane, and with a sole exception (alamethicin: see later) they are unable to form discrete lipid soluble complexes with simple inorganic ions. We are concerned with substances which can form a hole through which the ions can move.

GRAMICIDIN[27, 34]: FUNCTIONAL ASPECTS

At a structural and mechanistic level, the best described of the channel formers is gramicidin A. This substance induces ion permeability in liposomes,[28] red blood cells, [2, 29] mitochondria,[28, 30, 31] and electroplax[32] and renders phospholipid bilayers electrically conductive.[6, 27, 29] Unlike the situation with the neutral ionophores (valinomycin, nonactin), the induced

conductivity due to gramicidin is reported to show a square law dependence on the concentration of the antibiotic, [29, 33] so that

$$G = G_0(1 + k'[\text{val}]) \text{ but } G = G_0(1 + k''[\text{gram}]^2).$$

Here, G_0 is the conductance of the unmodified membrane and is exceedingly low (See Chapter 1, p. 34) and G is the measured conductance;
k' and k'' are proportionality constants. In spite of some doubt about the
reality of this observation (due to the difficulty of knowing precisely the
concentration of gramicidin in the bimolecular region of the model
membrane) there seems to be fairly general agreement from this and other

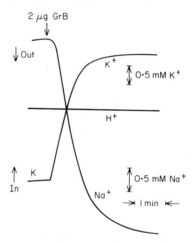

FIG. 16. Loss of internal Na^+ from dog red blood cells in exchange for external K^+,
mediated by gramicidin. The measurements were made with ion-specific glass
electrodes for Na^+, K^+ and H^+. (From ref. 2.)

evidence that two molecules of gramicidin are involved in the formation of
a conducting pathway (but see refs 27 and 34 for a cautious assessment of
this point).

The conductivity pathway is selective for monovalent cations, but
among these the degree of selectivity is small. Thus it enables the internal
Na^+ of dog red blood cells to exchange for extracellular K^+ (Fig. 16). The
selectivity sequence, provided by measurements of biionic potentials (see
p. 116) on lipid bilayer membranes is in the order[35]

$$H^+ > NH^+_4 > Cs^+ > Rb^+ > K^+ > Na^+ > Li^+.$$

In spite of the low degree of selectivity, the very order of this sequence is
worthy of comment since it is the same as the sequence of ionic mobilities
in water, and this, taken with other evidence, suggests that there may be a
continuous row of water molecules in the gramicidin channel.

Channels and Carriers: Some Distinguishing Features

Single-channel Conductance Events. An important feature of the gramicidin channel is that when ion conductance measurements are made in the
presence of absolutely minimal amounts of gramicidin it is possible to
detect discrete conductance events separated in time.[36] No such transient
events have ever been observed with mobile carrier substances even when
the means of detection have been refined to exclude all possibility of
electrical and mechanical noise, and the amplification increased to enable
the detection of conductance changes as low as 10^{-12} Ω^{-1}. As the concentration of gramicidin is increased, the separate conductance events
occur closer and closer together in time until the current fluctuations even
out and are no longer observable. The step height of the conductance
events is very uniform, the larger events being apparently integral multiples
of a unit event. The basic process then appears to be the opening and

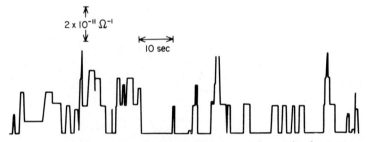

FIG. 17. Unitary conductance events in a bilayer membrane in the presence of a
very small amount of gramicidin. Note the uniformity of the basic step height.
Reproduced by permission of Hladky and Haydon (1972), *BBA*, **274**, 294.

closing of single and independent conductance channels. The time during
which the conducting channels remain open is continuously variable, and
this suggests that the opening and closing processes are random occur-
rences in time.

The means of recognising the unit event is also the means of measuring
the unit channel conductance, and hence the ion fluxes through single unit
channels. By comparison with the mobile carrier substances, the ion flux
in the gramicidin channel is found to be high: certainly much higher than
could be accounted for by the diffusion of a mobile ion-gramicidin complex
in the membrane.

Thermal Effects. A feature which highlights the differences between
channels and carriers is the thermal dependence of ion permeation in
treated lipid bilayers.[37] As the temperature of a bilayer treated with a neutral
carrier (such as valinomycin or nonactin) is lowered below the temperature
of the hydrocarbon phase transition the high induced conductivity state
is suddenly abolished. On lowering the temperature of a bilayer treated
with gramicidin there is no such discontinuity: only an increase in the

temperature dependence of the ionic conductance (see Fig. 18). Below the phase transition, the conductivity of the gramicidin-treated membrane becomes gradually and continuously less: above the phase transition

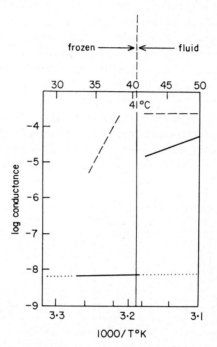

FIG. 18. The effect of temperature on the conductivity of a lipid bilayer treated with a mobile ion carrier (nonactin, ———) and a channel forming substance (gramicidin, - - - -). The membrane was composed of an equimolar mixture of diC 16:0 and diC 18:0 glycerols. The $\log_{10}$ conductivity is plotted against the reciprocal of the absolute temperature, i.e. in the form of an Arrhenius plot.

On cooling the membrane, freezing at 41° could be detected visually at the point when islands (or lenses) of thick lipid film supported by the black membrane became immobilised. This corresponded very closely to the temperature at which the nonactin-treated membrane ceased to conduct: the conductivity at this point fell by 4 orders of magnitude. The conductivity of the gramicidin-treated membrane shows no corresponding discontinuity: merely the development of a finite Arrhenius activation energy. (Adapted from ref. 37.) Copyright 1971 by the American Association for the Advancement of Science.

the conductivity is essentially independent of temperature. We conclude that the freezing of the hydrocarbon interior of the phospholipid bilayer prevents the free diffusion of the true ion carrier molecules, but has only a marginal effect (possibly by hindering flexibility) on the structured channel formed by gramicidin.

Dimensional Effects. Measurement of the current flow in unit conductance events reveals that whilst the unit channel conductance is insensitive to variations in the thickness of the membrane, both the duration of the unit channel and the frequency of opening increase in thinner membranes.[36] In these experiments, the thickness of the lipid bilayers was controlled both by varying the length of the hydrocarbon chains of the component lipid and by varying the chainlength of the hydrocarbon solvent used to dissolve the membrane forming lipid (as described in Chapter 1, p. 37). Membrane thickness was determined from capacitance measurements. In thinner membranes, the equilibrium shifts in favour of the conducting configuration. The constancy of the unit channel conductance leads to the suggestion that the structure of the coherent channel in the conducting mode is unaltered by varying the membrane thickness, and that thicker membranes may even be deformed ("dimpled") in the region of the conducting channel in order to allow it to span the barrier between the two aqueous interfaces. As the duration and frequency of channel opening increases down to the thinnest membranes (having a hydrocarbon thickness of about 26 Å), it is possible to set an upper limit to the length of the gramicidin channel, and if allowance for the thickness of the polar head groups is made, this comes to about 35 Å.

GRAMICIDIN: STRUCTURAL ASPECTS

The primary structure[38, 39] of gramicidin A is illustrated in Fig. 19. It is a peptide having fifteen residues, all of which have hydrophobic side-chain chemistry. The N-terminal is blocked with a formyl group and

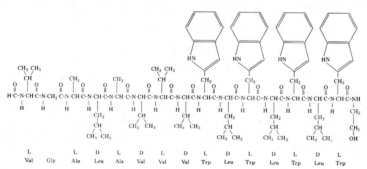

FIG. 19. The primary structure of gramicidin A.

the C-terminal is blocked with ethanolamine. The configuration of the aminoacids alternates L D L D . . . with the one exception of the optically inactive glycine, which may be said to replace a D residue.

Gramicidin forms stable dimers in non-polar solvents,[40, 41] and it is this feature, taken with the square law dependence of ionic conductance on gramicidin concentration in bilayers, which led to the idea that the coherent channel requires two molecules of gramicidin.[42] The problems

of assigning the three-dimensional structure of the gramicidin channel
are formidable; certainly far greater than for valinomycin which was discussed earlier in this chapter. With a trans-membrane channel such as
gramicidin, even the approach of studying stable configurations in non-

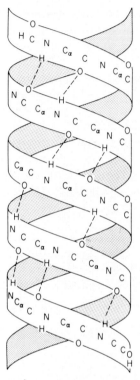

FIG. 20. A proposed structure for the gramicidin channel: the antiparallel-β
double helical dimer with seven residues per turn in the left-handed configuration.
Broken lines indicate hydrogen bonds. One N-terminal formyl group is visible
at the top left of the dimer and one C-terminal ethanolamine group is visible at
the bottom right. Reproduced with permission of Vetch et al. (1974), Biochemistry,
13, 5249.

polar solvents by spectroscopic methods is suspect, since it must be
evident that the environment provided by the bilayer is heterogeneous and
spacially oriented. In addition to this problem, it has been shown that in
non-polar solvents, gramicidin exists as an equilibrium mixture of four
interconvertible conformational species, all of which are probably dimeric.[41]
The equilibrium distribution of gramicidin can be "frozen" under carefully controlled conditions and this allows for separation of the conformers
by thin layer chromatography.

Spectroscopic examination of the isolated conformers has led to the proposal of a new family of parallel and anti-parallel hydrogen bonded double helical dimers. These are broad pitched structures which could accommodate a channel of sufficient dimensions for ion movement down the centre. A formal illustration of an antiparallel double helical structure of gramicidin is shown in Fig. 20. The inter-peptide hydrogen bonds are in the β arrangement (i.e. alternately pointing forwards and backwards in parallel with the axis of the helix). The amino acid side-chains (which are omitted from the diagram) extend outwards from the double helix to interact with the hydrophobic region of the bilayer, and the structure also ensures that the "channel" is lined with the oxygen atoms of the amino acid carboxyl groups.

In an earlier formulation of the gramicidin channel, dimerisation of the molecules (in single β helical arrangement) occurs by "head-to-head" interaction between the formyl residues at the N-terminal.[43, 44] Helices of this type, having six residues per turn, would have dimensions sufficient to span a phospholipid bilayer and to generate an internal channel of about 4 Å diameter. Although the physical basis for the head-to-head single helical dimer is less secure than for the double helical structures (the supporting spectral determinations having been made on the equilibrium mixture of gramicidin dimers rather than on isolated single conformers) the proposal is obviously attractive. The discontinuous nature of the conductance events are then explained by the momentary association of two gramicidin molecules anchored at their C-termini to opposing surfaces of the membrane, and independently mobile. As they come into confluence, a coherent channel is formed and this is closed off as they move apart.

An illustration of the head-to-head model of the gramicidin dimer is shown in Fig. 21.[42] As with the double helical dimers in β configuration, the channel is lined with the hydrophilic groups of the peptide bonds, while the exterior of the structure is exclusively hydrophobic. Whilst it is improbable that the head-to-head single helical dimer exists as a stable and unique entity in bulk phase hydrophobic solvents, it cannot as yet be eliminated by the newer proposals as a candidate for the structure of the actual gramicidin channel in membranes. Indeed, it could be remarked that there is no particular reason why the structure of the conducting configuration should enjoy any long-term stability, and the ephemeral nature of the conductance events even suggest that it doesn't. Presumably, spectral examination of gramicidin in the special heterogenous environment provided by the bilayer structure of membranes will be required to carry this problem further. This approach could be well developed by studies of the comformation of gramicidin included in a sonicated single bilayer liposome preparation of the type described on p. 17 of Chapter 1.

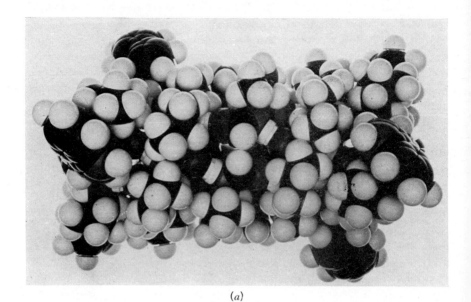

(a)

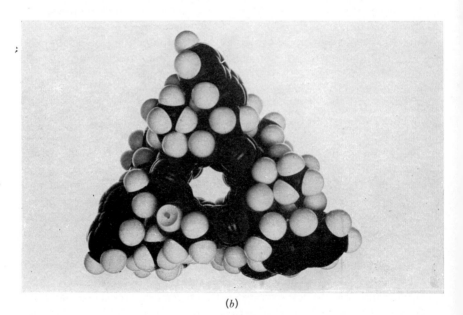

(b)

Fig. 21. Space-filling representation of the head-to-head model of the gramicidin dimeric channel. (a) Continuity of the helical arrangement in the region of union between the two N-terminal formyl groups at the centre of the dimer is apparent. (b) The head-to-head dimer seen end on: the internal channel has a diameter of about 4 Å. Reproduced by permission of Urry (1972), *BBA*, **265**, 113.

THE CYCLIC POLYENE ANTIBIOTICS

There is a large group of cyclic polyene antibiotics which exert their cytotoxic action by rendering the membrane of the target cell permeable to ions and other small molecules.[45] The most well known of these, filipin, amphotericin B and nystatin, are widely available and have been used

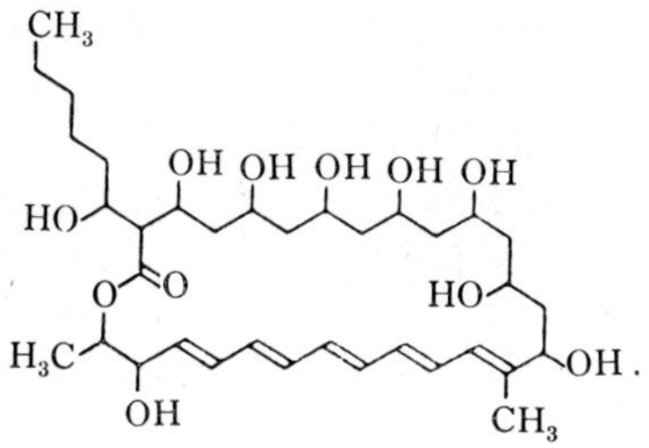

Filipin

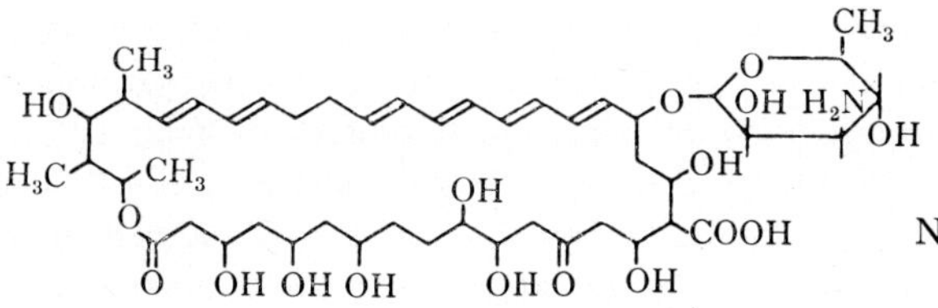

Nystatin

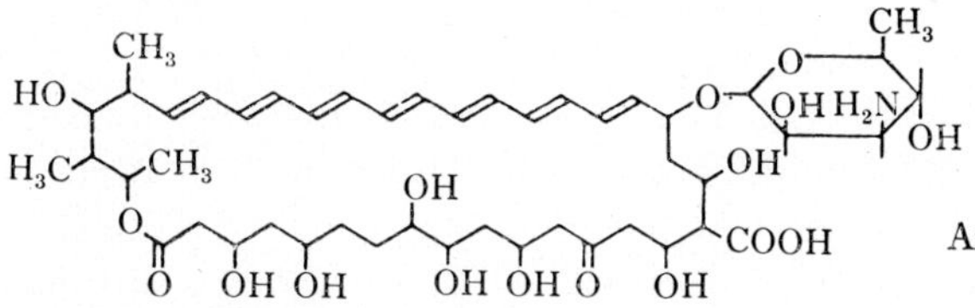

Amphotericin B

FIG. 22. The proposed structures of the polyene antibiotics, filipin, nystatin and amphotericin B.

in clinical medicine for the treatment of topical infections of fungal origin. The polyene antibiotics are inactive as cytotoxic agents against bacteria.[50]

They are all large lactone ring compounds containing a number of double bonds (hence "polyene"). Nystatin[46] has four conjugated double bonds, filipin[47] has five and amphotericin B[48] has seven. The rings also contain a hydrophilic zone due to the presence of an array of hydroxyl groups. Glycosyl and charged groups may also be attached (nystatin and amphotericin B) but filipin is neutral.

Cytolytic Properties

The cytotoxic effect of the polyenes may be simply measured and compared by their tendency to induce haemolysis of red blood cells. The order of potency is nystatin < amphotericin B < filipin.[49] The same order of potency is obtained by measuring the effects of the polyenes on the respiratory and glycolytic rates of yeast cells.[50]

The lytic action of nystatin and amphotericin B on red blood cells and fungal protoplasts may be prevented by the addition of a non-penetrating solute such as sucrose to increase the osmotic pressure of the solution in which the cells are suspended.[49, 51] Such protection against lysis does nothing to inhibit the immediate loss of intracellular potassium which occurs when the antibiotics are added to red cells; it merely prevents the terminal lytic event. Cells treated with filipin cannot be protected by the addition of non-penetrating solutes even when these are of macromolecular dimensions, and the reason for this is probably that the filipin-pore is so much larger than the pores created by nystatin and amphotericin B, and allows direct leakage of intracellular proteins. Following treatment of susceptible cells with filipin there is evidence of extensive membrane damage and fragmentation[52, 53, 54] which can be seen by phase-contrast, freeze-etch and negative strain electron microscopy. Such damage is absent from cells which have been lysed with nystatin or amphotericin B. These allow the ionic gradients of Na^+ and K^+ across the cell membrane (which normally are maintained by metabolic processes against a controlled, slow rate of leakage) to dissipate resulting in a net accumulation of ionic material within the cell and consequent osmotic lysis.[50, 51]

Effects in Model Membranes

It was by the use of model membrane systems that the insensitivity of bacteria towards attack by the polyene antibiotics could be understood.

(a) *Monolayers*. Firstly, in monolayer investigation,[55, 56] it was shown that whilst the polyenes cause an increase in the surface pressure of a layer formed from a total extract of red cell lipids, there is no measurable effect when the monolayer is formed from the phospholipids alone. Similarly, there is no effect of the polyenes on the surface pressure of a layer formed from bacterial lipids. The chief difference between the sensitive and insensitive preparations is of course the presence of sterols in the total lipid extract of red blood cells. Sterols are not found as a component of bacterial cell membranes and these are known to be insensitive to the cytotoxic action of the polyene antibiotics.[50, 57] The realisation of a requirement for sterols has been extended to investigations of the precise structural features on the sterol molecule which are required to confer sensitivity towards the polyene antibiotics[72, 73] and it should come as no surprise to readers of Chapter 1 to learn that only sterols having a planar nucleus, an aliphatic side chain at C_{17} and a β-hydroxyl group in the 3-position

(Fig. 3, Chapter 1) are effective. When applied to lipid monolayers at the air-water interface, the effect of the polyenes is greatest when the lipid consists of sterol alone: the presence of phospholipid actually reduces the expansion effect due to the polyenes. When pure sterol monolayers are used, then one may distinguish a clear order of potency for the effect of the polyenes in enhancing the surface pressure: nystatin < amphotericin B < filipin. The structural criteria for the biological activity and the characteristic physical interactions of the polyenes with membrane structures are also fairly precise. For example, perhydro-filipin (which lacks the conjugated double bond system of the parent substance) is inactive as a lytic agent and exhibits no interactions with cholesterol. Similarly, hydrolysis of the lactone ring of amphotericin B results in a loss of activity.[61]

(b) *Bilayers*. The polyenes are able to increase the permeability of cholesterol containing bilayers to ions, water and small hydrophilic solutes, by 5–6 orders of magnitude.[60, 61, 62, 63, 64, 65] Generally, studies with bilayers have been restricted to nystatin and amphotericin B as filipin reduces the stability of the bilayer film.[65] In contrast with the mobile current-carrying ionophores (valinomycin, nonactin, etc.), the increase in permeability varies as a high and variable power of the polyene concentration (4·5–10, depending on the composition of the membrane[60, 61, 65] (see Fig. 23)), and this suggests that a number of polyene molecules react together to form a conducting pathway. The electrical conductance of polyene treated membranes is not restricted to cations, and under some circumstances it is possible to regulate the ionic preference making it cation or anion selective.[67] There is a strict proportionality between the magnitude of the induced ionic conductance and the induced permeability to water and non-electrolytes, and on this basis, it is believed that there exists a common pathway in polyene treated membranes for all permeant substances.[63]

It will be recalled (Chapter 1, p. 40 *et seq.*) that osmotic and tracer measurements yielding the permeability coefficients P_{d_w} and P_{f_w}, and P_{d_i} and σ_i can be used to diagnose the mechanism of solute and water permeation through membranes. The data for water flux through untreated and polyene treated bilayers is recorded in Table 1. From the data in Table 1 it can be seen that whilst the diffusion flux rates are rather insensitive to the presence or absence of nystatin or amphotericin B, the filtration permeability of water increases and it is clear that $P_{f_w} > P_{d_w}$ for the anti-

TABLE 1. Water permeability measurements made on treated membranes having resistance of 100Ω cm² in 0·1 M NaCl.

	P_{f_w} cm sec⁻¹.10⁻³	P_{d_w} cm sec⁻¹.10⁻³	P_{f_w}/P_{d_w}	reference
untreated	1·14	1·06	1·1	68
nystatin	4·0	1·2	3·3	63
amphotericin B	1·8	0·6	3·0	63

biotic treated membranes. This indicates that there is a cooperative mechanism of water flow, whereby water molecules interact with each other in a hydrated pathway is in contrast with the situation in untreated membranes (in which $P_{f_w} \simeq P_{d_w}$) where water movement occurs by a diffusion mechanism.[68]

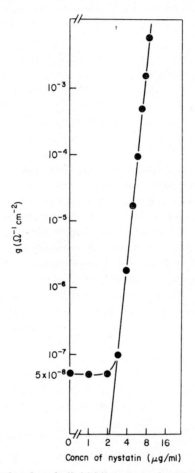

FIG. 23. Conductance of a phospholipid bilayer as a function of nystatin added at equal concentrations on either side (From ref. 65.) Reproduced by permission of Finkelstein and Cass (1967), *J. Gen. Physiol.* **52**, 145, suppl.

The polyenes produce a graded effect dependent on solute size, on the measured filtration rate of water through treated membranes.[64] This also is in contrast to the situation of untreated membranes, across which all the solutes (of Fig. 24) are impermeant and for which a true limiting value for the filtration permeability of water can be obtained. Only with the larger, impermeant solutes which still give a value of unity for the reflection coefficient can the limiting filtration rate of water flow in treated membranes be determined. With the smaller solutes ($\sigma < 1$) the counter-flow of the solute is sufficient to prevent the limiting expression of water flow through the polyene induced pathway. There is an apparent sharp cut-off between

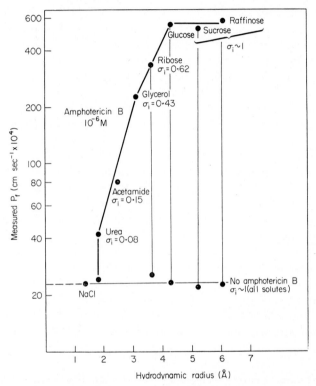

FIG. 24. The relationship between the filtration permeability of a phospholipid bilayer treated with amphotericin B and the effective hydrodynamic radius of a number of solutes. The measured values of P_f reach an upper limit of 549 cm sec^{-1} × 10^4 in the presence of impermeant solutes such as raffinose (a trisaccharide). The reflection coefficients σ_i, for the permeant solutes increase monotonically with molecular size. In the absence of the polyene antibiotic, $\sigma_i \sim 1$ for all solutes, and $P_{f_w} \sim P_{d_w}$ indicating diffusional flow. (From ref. 64.) Reproduced by permission of Andreoli et al. (1969), J. Gen. Physiol. **53**, 133.

the pentose and the hexose sugars (though the permeability of the treated membranes to glucose and even sucrose does increase slightly) and it is therefore likely that the effective pore radius is at least 5·5 Å in diameter.

(c) *Liposomes.* As was pointed out earlier, it has not proved feasible to make measurements on bilayers after treatment with filipin.[66] This is because the bilayer membranes containing cholesterol become unstable and break. In considering the nature of membrane lesions induced by filipin, we should recall the inability of macromolecular solutes (such as haemoglobin and high molecular weight dextrans) to prevent leakage of intracellular contents (p. 140). Such phenomena could be explained on the basis of the existence of pathways having dimensions sufficient to be discerned by electron microscopy.

The technique used to visualise membrane pores is the same as that used to demonstrate the multilamellar structure of the liposomes—negative stain electron microscopy. The negative stain (generally phosphotungstate) is understood to penetrate into all the water accessible spaces of the preparation so that water-filled pores will take up the stain and will appear electron dense if the dimensions of the pore are sufficient to accommodate the negative stain molecule. The water-immiscible membrane components are unstained and show up against the dense stained aqueous phase.

In this way it has been possible to demonstrate staining features ("pits") in susceptible cell membranes treated with filipin[54] which might be the actual pores. These have a radius between 40 and 75 Å—quite large enough to allow leakage of the main intracellular proteins. The lesions bear a striking resemblance to the dark staining regions present in negative stain electron micrographs of membranes from cells which have been lysed with antibody and complement (see Chapter 4). Negative stain electron microscopy of liposomes following treatment with filipin also reveals the presence of stained pits—but here we add the necessary proviso that cholesterol must be a component of the lipid mixture used to generate the liposomes. It should be remarked, however, that there is no evidence that these dark staining pits are true holes which traverse the membrane; the resolution of the staining technique is most probably insufficient to allow this conclusion to be drawn directly. The pits do surely represent structural features which are closely related to the fundamental lesion in the membrane induced by filipin. Freeze fracture electron microscopy, which clearly reveals the membrane lesions within the fracture plane of cholesterol containing membranes from *Acholeplasma laidlawii*, red cells and liposomes (see Fig. 26) certainly provides no support for the idea of a coherent structural pathway penetrating right through the membrane, although the filipin generated lesions are clear for all to see.[53]

The permeability properties of nystatin and amphotericin-B treated liposomes and cells in suspension have been explored by the techniques of light-scattering and release of internal markers (see Chapter 1, p. 29).[59] A requirement for the presence of suitable sterols has been discerned both for liposomes and for the cells of *Acholeplasma laidlawii*, which can incorporate defined sterols provided in the growth medium.[69, 70] In order to be effective, sterols must be of the type which interact with phospholipids to cause film compression [71] and phase buffering (see Chapter 1), i.e. they must be like cholesterol.[58]

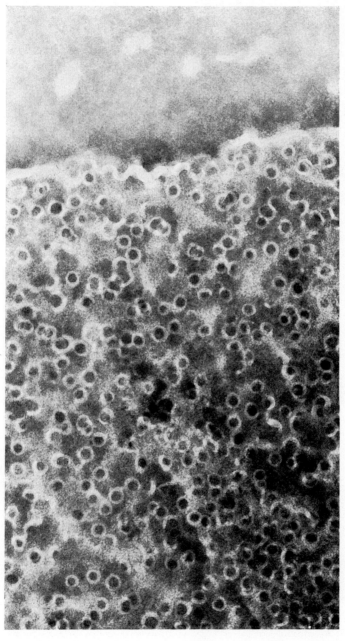

FIG. 25. Negative stain electron microscope photograph of lesions induced in red cell membranes by filipin. These lesions bear a resemblance to those created by the action of antibody + complement; see p. 185, Chapter 4. Reproduced by permission of Kinsky *et al.* (1967), *BBA*, **135**, 844.

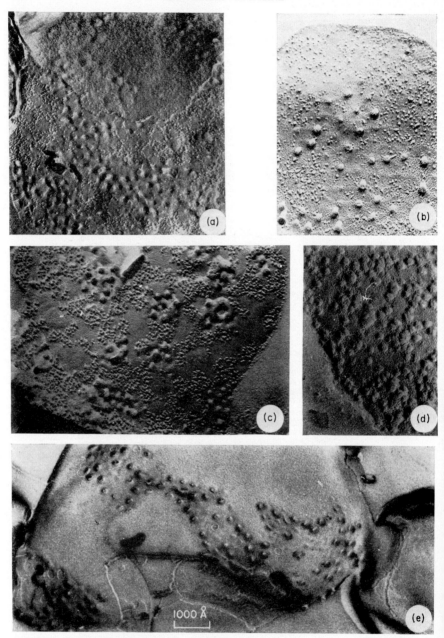

FIG. 26. Freeze fracture and etched images of membranes treated with filipin.

(a) and (b) Membranes from *Acholeplasma laidlawii* grown in a medium containing cholesterol. (a) Convex freeze etch image revealing protoplasmic fracture surface PF and outer membrane surface ES. (b) External fracture face EF.

Caption continued on facing page.

Amphotericin B and nystatin allow the passage of small molecules like the alkali metal cations, tetrose and pentose sugars and urea, the rate of flux depending on the size of the molecule (see Fig. 27).[59] The rate of flux of the alkali metal cations is in the order $Rb^+ > K^+ > Na^+ > Li^+$. This is the order of increasing size of the hydrated ionic radii and provides further evidence for the idea of water-filled pores for the transport of solutes. Larger molecules such as enzyme proteins, nicotinamide coenzymes ($NADP^+$), glucose-6-phosphate, sucrose and alkaline earth cations remain impermeant. Cholesterol-containing liposomes modified with amphotericin B are just permeated by glucose, and this sets an upper limit on the dimensions of the pore (see Fig. 28): the nystatin pore is not permeated by glucose and must therefore be smaller.

Formation of Polyene-Cholesterol Complexes

Some insight into the mode of action of the polyenes in inducing pores in cholesterol-containing membranes has been gained from calorimetric studies.[72, 73, 74] It will be recalled (Chapter 1, p. 25) that the effects of cholesterol on the thermal phase properties of phospholipid membranes can be explained by a reduction in the energy content of the phase transitions which are observed as the hydrocarbon chains melt (or freeze). In a membrane (or liposome) composed of a single phospholipid species, the phase transitions are well defined and have a considerable energy content, but as the cholesterol content is increased towards 50%, the heat absorption due to the phase transition disappears altogether. This is the condition of free cholesterol within the membrane, randomly distributed in the plane and causing maximum disruption to the ordered array of phospholipid hydrocarbon chains, which are thereby hindered from reacting in a cooperative manner.

Suppose that now, the cholesterol is gathered up into a number of

---

(c) and (d) Human red cell membranes. (c) protoplasmic fracture face PF; (d) outer membrane surface PS.

(e) PC-cholesterol liposomes, fracture face.

In the fracture planes of the cell membranes large aggregates and excavations (150–250 Å in diameter) can be seen in addition to the normal 80 Å intercalated particles. The large aggregates are mainly associated with the external fracture surface EF, and the excavations with the protoplasmic fracture face PF; this can most clearly be seen in the images of *Acholeplasma* membranes in (a) and (b) above. Swellings on the outer etch (PS) surfaces of Acholeplasma and red cell membranes due to the presence of large aggregates within the bilayer can also be seen.

Treatment of cholesterol-containing liposomes also results in the presence of large aggregates contained within the fracture planes of the outermost leaflets.

No effects of filipin can be seen by freeze fracture electron microscopy of cell membranes or liposomes which lack cholesterol.

(From ref. 53.)

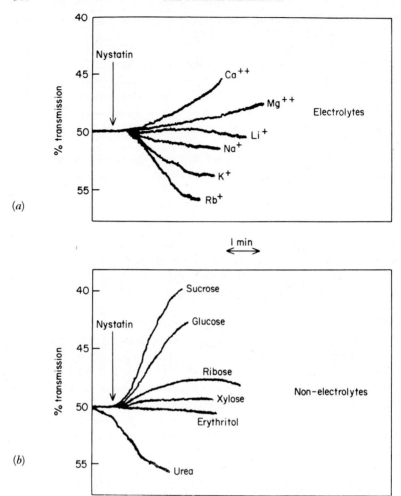

Fig. 27. The effect of nystatin on the light-scattering properties of *Acholeplasma laidlawii* (grown with cholesterol) suspended in various isotonic media. (*a*) Among salts, the fastest rate of swelling is induced by penetration of Rb⁺ (Cs⁺ not tested). Ca^{2+} and Mg^{2+} cannot pass through the antibiotic treated membrane, and the increase in light-scattering (due to cell shrinkage) is caused by loss of permeant materials out of the cells through the pathway generated by the nystatin. (*b*) Among non-electrolytes, only urea and erythritol appear to pass through the antibiotic modified membrane. (From ref. 59.) Reproduced by permission of de Kruijff *et al.* (1974), *BBA*, **339**, 30.

restricted locations in the membrane by complexation with an added component: this would leave areas of the membrane free of cholesterol, and would allow the phospholipid molecules in these regions to realign, and to react in a cooperative manner once again. The defined requirement for cholesterol in polyene antibiotic action suggested that complex formation between these substances within the membrane could occur. One result of such complex formation would be (of course) to clear regions of the membrane free of cholesterol. The question could be asked: does complex formation between polyenes and cholesterol allow the re-expression of the energy absorptive phase transitions through the re-attainment of structural alignment of the phospholipid hydrocarbon chains? The proposition has

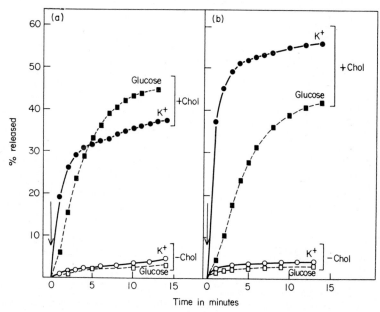

FIG. 28. Effect of filipin (a) and amphotericin B (b) on the release of K^+ and glucose from liposomes prepared with and without 16 moles % of cholesterol. The appearance of K^+ externally was measured with a K^+ specific glass electrode, and external glucose was measured spectrophotometrically (the method is described on p. 188, Chapter 4). After treatment with filipin, glucose and K^+ both leak from cholesterol-containing liposomes at a similar rate. After treatment with amphotericin, the rate of loss of K^+ is somewhat faster than the loss of glucose. There is no effect of the polyenes on liposomes prepared without cholesterol. Reproduced by permission of de Kruijff et al. (1974), BBA, 339, 30.

been tested both for liposomes[72, 73] and for the membranes of Acholeplasma laidlawii,[70] in which the cholesterol and hydrocarbon components can be controlled through regulation of the growth medium.

In both systems the effect of cholesterol is to reduce the energy content of the thermal phase transition as measured by integrating the peak obtained by differential scanning calorimetry (see p. 18, Chapter 1). Addition of the polyene antibiotics results in an increase in the energy content of the phase transition suggesting that the cholesterol is no longer available to disrupt the organisation of aligned hydrocarbon chains by its random distribution in the plane of the membrane.[72] This is clear evidence for a close asociation between the cholesterol and the antibiotic molecules. Under some circumstances, it is possible for the polyenes to complex all of the cholesterol present in the *A. laidlawii* membranes.

Complex formation between cholesterol and the polyenes can also be observed by optical means, since the polyene antibiotics have well defined

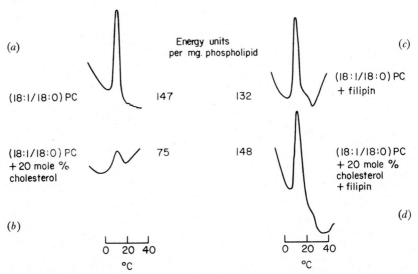

FIG. 29. Differential scanning calorimetry heating curves to illustrate the restoration of the crystalline to liquid-crystalline phase transition in PC-cholesterol liposomes by the addition of filipin. (*a*) Illustrates the normal heating curve for 18:1/18:0 PC. The energy of the phase transition (obtained by integrating the peak) is 147 energy units/mg phospholipid. (*b*) The presence of 20 moles % of cholesterol reduces the energy input at the phase change to half. (*c*) Filipin added to simple phospholipid liposomes having no cholesterol has no effect on the thermal phase properties of the preparation. (*d*) Filipin added to liposomes containing 20 moles % of cholesterol results in total restoration of the thermal phase properties of the simple phospholipid system. Adapted from Norman *et al.* (1974), *J. Biol. Chem.* **247**, 191.

and suitably responsive absorption spectra.[72, 73] By this means, it can be shown that complex formation between filipin and the cholesterol of *A. laidlawii* membranes occurs before the onset of enhanced permeability to potassium, and so it would appear that complex formation is a necessary precursor of pore formation. The complexes formed between polyenes and microcrystalline dispersions of cholesterol in water are most stable at low temperatures, suggesting that they are stabilised by "hydrophobic" forces of interaction; however the onset of enhanced permeability in cells at low temperatures is generally delayed. The picture which emerges is one of a

primary interaction (approximately 1:1) between the polyene and the
cholesterol which is initially distributed in a random manner throughout
the membrane. These elementary wedge-shaped complexes (see Fig. 30)
then diffuse laterally to form larger structured clusters which ultimately
constitute the actual pores in the case of nystatin and amphotericin B.
It is this process of lateral diffusion which is diminished at lower temperatures due to the sluggish thermal motion within the hydrocarbon region,

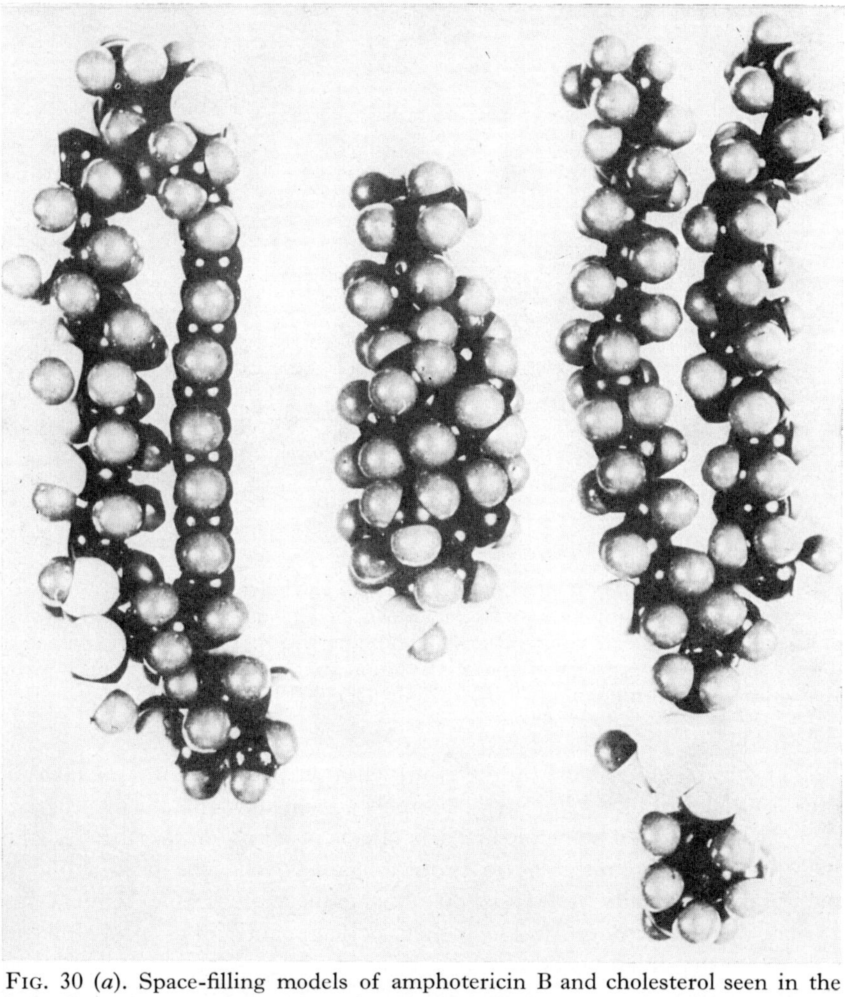

Amphotericin B   Cholesterol   Phosphatidyl choline

FIG. 30 (a). Space-filling models of amphotericin B and cholesterol seen in the
long dimension together with a phosphatidyl choline molecule. With the "head
groups" placed in line, it can be seen that the length of PC molecule and the polyene
are similar; in no way could the polyene span a bilayer of two PC molecules thickness. The rigid-bonded straight arm of the polyene, and the hydrophobic portion
of cholesterol are similar in length.

F

which of necessity is affected by the presence of cholesterol. In the case of filipin massive aggregates (150–250 Å diameter) are formed within the bilayer; and it is these which are observed by freeze fracture electron microscopy of both cell membranes and cholesterol-containing liposomes.[53]

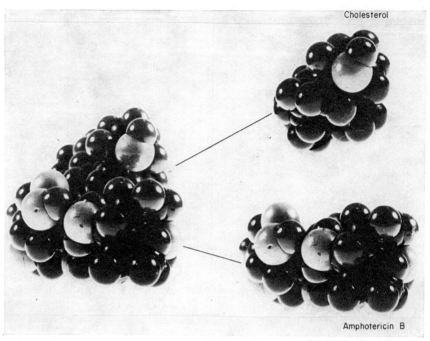

FIG. 30 (*b*). Space-filling models of amphotericin B, cholesterol and their elementary wedge-shaped complex, seen looking down the long axes from the 3-OH group of cholesterol and the mycosamine ring of amphotericin B. The carboxyl group and one of the hydroxyl groups of the mycosamine ring of the polyene can be seen. Reproduced by permission of de Kruiff and Denel (1974), *BBA*, **339**, 57.

Rather than forming a pore through the membrane, current thinking favours the idea that the filipin-cholesterol complex actually disrupts the membrane. A reconsideration of some of the basic effects of filipin on membranes (the inability of macromolecules to protect against lysis; the production of membrane fragments; the tendency to make bilayers unstable) renders this a plausible description.[75] The other polyene antibiotics probably do generate water-filled pores.

A Molecular Model for the Transmembrane Pore
Proposals for the structure of the transmembrane pore of amphotericin B and nystatin in cholesterol-containing membranes have been made.[67, 73-77]

These are based upon consideration of both membrane phenomena as outlined above, and the fitting of space-filling models of these molecules with cholesterol.

Examination of the structures (see Fig 22) of nystatin and amphotericin B shows that these substances are amphipathic: along one side there is an extensive hydrophobic region, dominated by the conjugated and rigid array of double bonds, and broken only by a single hydroxyl group at the "tail" (where it is almost enclosed by three methyl groups). Along the other side of the loop there is an arrangement of keto and hydroxyl groups. At the "head end" of the molecule two ionisable groups, caboxyl and amino are found as components of a mycosamine ring (i.e. 3,6-dideoxy-3-amino-D-mannose).

It is presumed that the elementary 1:1 complexes between the polyene and cholesterol (Fig. 30) are initially formed by interactions between the rigid nucleus of the cholesterol and the rigid π-bonded side of the anti-

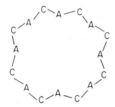

FIG. 31. Schematic arrangement of the hydrophilic well, composed of eight cholesterol-amphotericin B elementary complexes. Reproduced by permission of de Kruiff and Denel (1974), *BBA*, **339**, 57.

biotic—such a complex might be expected to be much more stable than any complex between the flexible hydrocarbon chains of the phospholipid and the cholesterol. There is no contact between the hydrophilic groups of the two molecules and the forces of stabilisation are exclusively hydrophobic. Because elementary complex formation in this way involves the introduction of the hydrophilic aspect of the polyene into the hydrophobic region of the bilayer, the alignment of the complex normal to the plane of the bilayer is intrinsically unstable. However, it is possible to form complexes between cholesterol and the polyene by attack on either side of the elongated loop of the antibiotic, and in this way, a number of the elementary complexes may aggregate to form a hydrophilic well which penetrates halfway through the bilayer (see Figs 31 and 32). Consideration of the angular and spacial dimensions of the polyene-cholesterol aggregate leads to a circular arrangement of eight amphotericin-cholesterol units having the formal structure $(—A—C—)_8$.

The aggregate is anchored to the surface of the membrane by the β-hydroxyl groups on the 3-positions of the cholesterol molecules, and the polar groups (carboxyl and amino) attached to the "head" of the polyene.

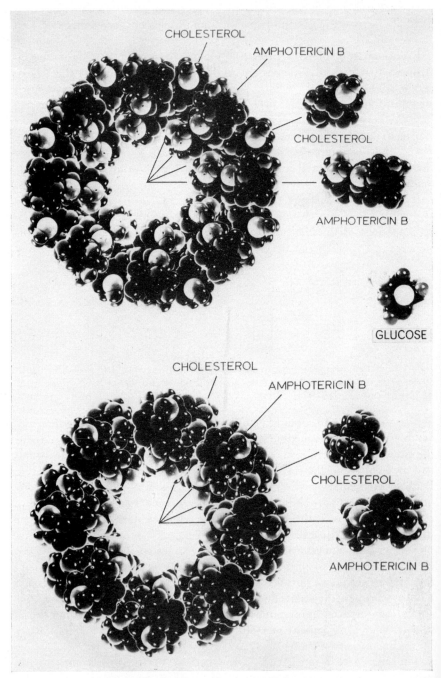

FIG. 32. *Caption at bottom of facing page.*

With the polar head of the polyene lying coplanar with the phosphate groups of the bilayer, the elongated loop will reach approximately into the region of the terminal methyl groups of the lipid hydrocarbons. The hydrophobic

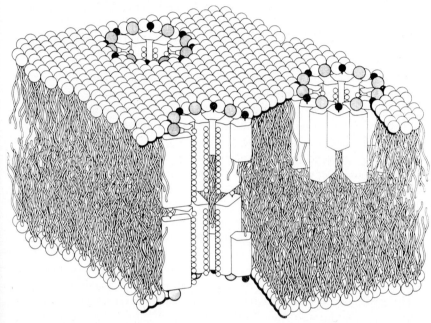

FIG. 33. Schematic representation of a well and a complete pore formed by amphotericin B and cholesterol in a phospholipid bilayer. The conducting pore is formed by the end-to-end union of two wells or half pores. (From ref. 75.) Reproduced by permission of de Kruiff and Denel (1974), *BBA*, **339**, 57.

zones of the cholesterol-polyene complexes are presented outwards from the aggregate, to interact in the hydrophobic region of the bilayer. The extended arrays of hydroxyl and carbonyl groups of the polyene rings are turned inwards as the lining of a hydrophilic cylinder. Two such cylinders anchored to opposite sides of the bilayer are of course necessary to form a transmembrane pore, and it is thought possible that the lone hydroxyl groups at the "tail" of the polyene may be involved in the anchoring of two

FIG. 32. Space-filling models of the (—A—C—)$_8$ well (half pore). (*Top*) Looking down the well from the head group region. The 3-OH groups of cholesterol are clearly visible near the periphery, and the carboxyl groups of the polyene can be seen lining the entrance to the well. At the lower right of the figure there is a space filling model of a glucose molecule, which can be seen to have about the right dimensions to fit the pore of the (—A—C—)$_8$ model for amphotericin B. (*Bottom*) Looking up from the bottom of the well. The hydrophilic lining of the well can be discerned. (From ref. 75.) Reproduced by permission of de Kruiff and Denel (1974), *BBA*, **339**, 57.

half-pores together (see Fig. 33). The diameter of the hydrophilic pore so generated for amphotericin B is 8 Å, and which is just sufficient to accommodate a molecule having the dimensions of glucose.

INDUCING ELECTRICAL EXCITABILITY

In all situations of induced ionic conductivity so far considered there has been an adherance to Ohm's law, $V = IR$. R, the resistance of the membrane is initially very high; by application of the neutral ionophores, and the channel forming polyenes and gramicidin, a new reduced value of R is achieved which then remains essentially constant. Certainly in none of these situations of induced ionic permeability has R varied as a high power of the membrane potential, which is the necessary condition for the generation of excitability in a membrane. Excitable membranes are found widely distributed in nature. As well as being the dominant functional characteristic of the nerve and muscle membrane, excitability has been described in epithelia,[78] plant cells[79] and also in a number of primitive algae,[80] protozoa[81] and ova.[82, 83] The generation of action potentials in an excitable membrane is a further development, requiring defined ionic gradients and an appopritate sequence of conductance changes.

The prospect of induced excitability in inert structures was illustrated by Mueller, Rudin and their colleagues along with the very first description of the phospholipid bilayer as a model of a biological membrane.[84] They treated the bilayers with "excitability inducing material", EIM—which was present as an impurity in batches of egg albumin which they were using to study the interaction of proteins with their model membranes. EIM has since been generated from cultures of *Aerobacter cloacae* and extensively purified.[85] It is a highly aggregated ribonucleoprotein of complex multisubunit structure, quite insoluble in both aqueous and hydrocarbon solvents. Other substances capable of inducing excitability in phospholipid bilayers include the small molecules, monazomycin, alamethicin and DJ400B, and of these, good structural information is now available for the last two.

ALAMETHICIN

Alamethicin is a polypeptide of 18 amino acid residues isolated from cultures of *Trichoderma viride*. Until recently it was thought to have the cyclic structure illustrated in Fig. 34 (a),[86] but new spectral evidence (NMR)[87] suggests that it is in fact an open chain structure, having N-acetyl methylalanine and B-phenylalaninol as N- and C-terminal groups (Fig. 34 (b)). DJ400B is a cyclic polyene[88] (Fig. 34 (c)), and monazomycin (MW 1400)

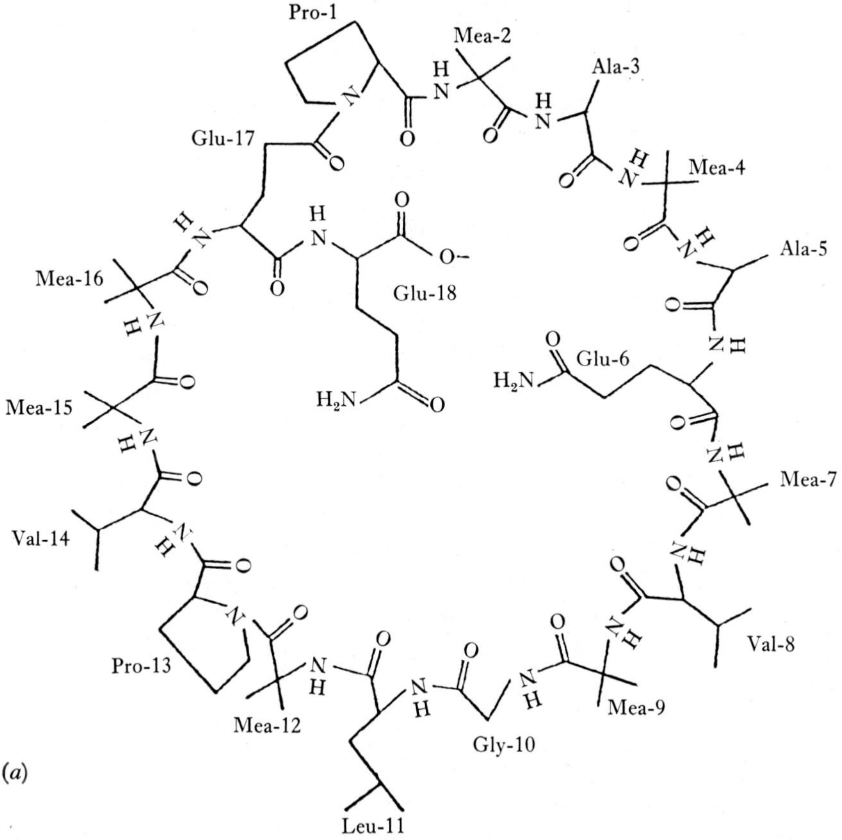

(a)

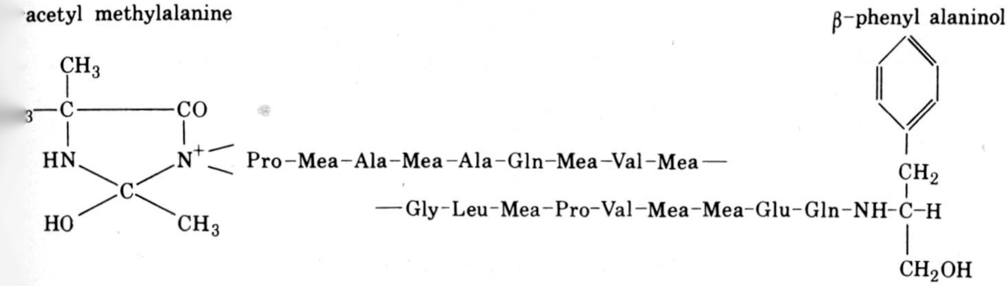

acetyl methylalanine

β–phenyl alaninol

Pro—Mea–Ala–Mea–Ala–Gln–Mea–Val–Mea—

—Gly–Leu–Mea–Pro–Val–Mea–Mea–Glu–Gln–NH–C–H

FIG. 34 (a) and (b). Cyclic and linear structures for alamethicin. The sequence of amino acids in both proposals is the same. N.B. Mea is α-methyl alanine (i.e. α-aminoisobutyric acid). (From ref. 87.)

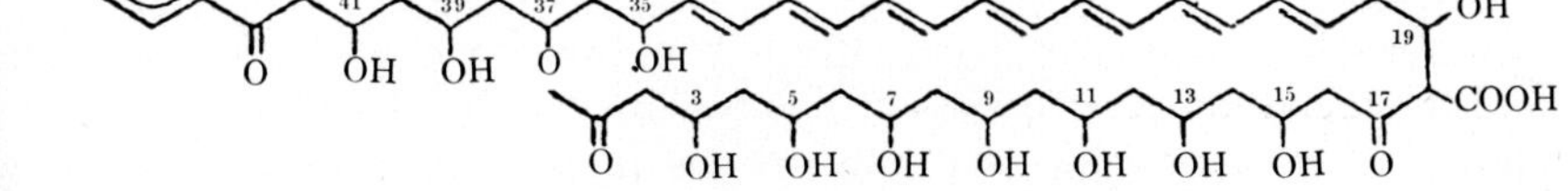

FIG. 34 (c). Structure of DJ400B. (From ref. 88.)

a *Streptomyces* metabolite may also be related to the polyene antibiotics discussed earlier.

Steady-state Conductance Properties

In many ways the transient and the steady-state conductance properties of bilayers treated with these substances resemble closely those which are found in cell membranes[89, 90] and may be described quantitatively by the equations of Hodgkin and Huxley, which were originally formulated as a mathematical description of conductance changes which occur in nerve and muscle following depolarisation.

The conductance of an alamethicin-treated bilayer at zero applied potential is very low—only about one order of magnitude above the conductance of the untreated bilayer.[91] As the potential across the membrane is increased, the conductance increases dramatically, as can be seen in Fig 35. It is clear that the current-voltage relationship is non-linear (non-Ohmic), with the current rising very steeply when the applied voltage exceeds about 60 mV. The asymmetry about the zero of applied potential arises from the fact that the alamethicin was applied to one side of the membrane only; over extended periods some leakage across the membrane can occur and this reduces the effect of alamethicin as a current rectifier. The non-linearity and the steepness of the current-voltage curve suggests that there are two conducting states, one of which is highly dependent on the applied voltage.

In a similar manner to the polyene antibiotics, the increase in conductance varies as a high power of the antibiotic concentration (Fig. 36 (*a*)), again indicating an extensive degree of molecular association in the formation of a conducting pathway. In the case of alamethicin, the conductance also varies by the same power factor of the applied voltage and of the metal ion concentration (Fig. 36 (*b*) and (*c*)),[92, 93] so that the conductance varies according to

$$G = \text{const} \times [A]^n [M]^n \left[1 + k. \exp \frac{nzeV}{kT} \right],$$

where G is the conductance, A is the concentration of alamethicin in the bilayer, M is the concentration of the permeant ions, z is the valency of the cation, V is the voltage, e is the electronic charge and k is Boltzmann's constant.

The exponent n, which expresses the degree of association, is variable (lying in the range 5–10 for alamethicin) and depends on a number of factors including the composition of the membrane lipids. (*N.B.* In the hands of some investigators,[94] separate exponents appear to be needed to describe the dependence of conductance on the alamethicin concentration and the metal ion concentration. The reasons for this discrepancy are not clear, but it is possible that the various ingredients used to form the bilayers in different laboratories could be a contributory factor. One of the problems here is deciding exactly what is the effective concentration of the alamethicin in the bilayer.[95] Whilst this must be recognised as a serious and complicating factor, the discussion here is developed on the basis of there

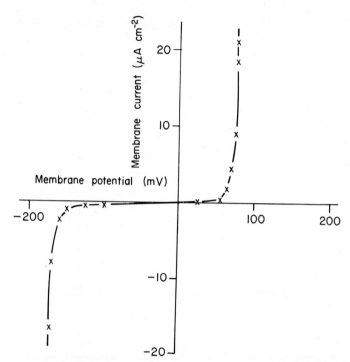

FIG. 35. A graph relating membrane current to membrane potential ("*I-V*" curve) for a phospholipid bilayer in the presence of alamethicin which was added at a concentration of 10^{-7} g/ml to the aqueous phase on one side of the membrane only. The salt concentration (NaCl 0·1 M + histidine 0·005 M, pH 7·0) was the same on both sides of the membrane. The slope of the curve at the zero of applied potential is very slight, showing that the membrane has a high resistance ($R = V/I$). When the applied potential exceeds $+60$ mV, the membrane resistance suddenly drops to a very low value. The asymmetry about the zero of applied potential is due to the fact that the alamethicin was added to one side of the membrane only—the side to which the positive potential was applied. Some leakage of alamethicin has occurred, and this accounts for the reverse current which flows when the potential exceeds -150 mV. (From ref. 91.)

being a common exponent relating the conductivity to the concentration of alamethicin and the salt concentration, and to the applied potential.)

A membrane potential due to an ion concentration gradient is sufficient to drive the alamethicin into its conducting configuration,[91] as illustrated in Fig. 37(*a*), where NaCl (0·05 M), present on one side of the membrane only, produces a resting potential of 17 mV: the polarity of the potential indicates that it derives primarily from a flow of cations down a conducting pathway. By treating the membrane with a polycation (such as histone or protamine) and so reversing the sign of the membrane surface potential, the selectivity of the alamethicin pathway alters to discriminate in favour of anions. At the same time the current-voltage curve is shifted so that the conducting pathway is now open at zero applied potential (Fig. 37(b)).

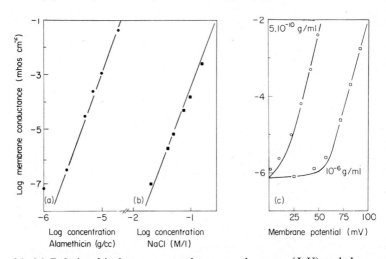

FIG. 36. (a) Relationship between membrane conductance (I-V) and the concentration of alamethicin added to the salt solution (NaCl 0·1 M) bathing a phospholipid bilayer. The potential across the membrane was kept constant at 60 mV. (b) Relationship between membrane conductance (I-V) and the concentration of salt added to the aqueous phase bathing a phospholipid bilayer which had been treated with alamethicin (10^{-5} g/ml). The potential was kept constant at 60 mV. (c) The relationship between membrane conductance (I-V) and the applied potential at two different concentrations of alamethicin in the aqueous phase. The electrolyte in both aqueous compartments was NaCl (0·1 M). Notice that the graphs relating log conductance with the log concentrations of alamethicin (at constant potential, and NaCl (a)) and the log concentration of NaCl (at constant potential and alamethicin concentration (b)) and the applied potential (at constant alamethicin and NaCl (c)) have a common slope. For this particular membrane (composed of sphingomyelin and α-tocopherol) the slope of these graphs is about 6. Reproduced by permission of Mueller and Rudin (1968), *Nature*, **217**, 713.

When a rectangular current pulse is applied to an alamethicin-treated bilayer in its normal low conductance resting state (no ion gradient) the voltage immediately rises to a high level[91] (Fig. 38) but this now has the effect of switching the alamethicin into a high conductance configuration, and as a result the voltage drops exponentially in spite of current flow being maintained, until the low conductance configuration is reasserted. With care, and by increasing the depolarising current to a critical level, one may set off a train of rhythmic oscillations (Fig. 39).

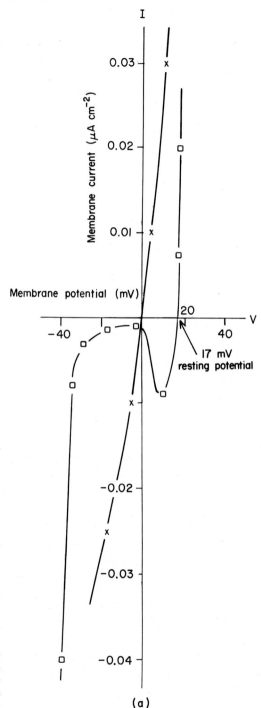

FIG. 37 (a). Current-voltage (I-V) curve for an alamethicin treated bilayer, having NaCl (0·05 M) on one side of the membrane only (□——□.) There is now a 'resting' potential of 17 mV at the point of zero current flow, which is caused by the ionic concentration gradient, and the system is in its conducting mode at this point (cf. Fig. 35). The ion gradient also causes the negative resistance phenomenon, whereby the current flow becomes increasingly negative as the potential is raised between zero and 10 mV.

The I-V curve of Fig. 35 has been included for comparison (×——×), but note that the scale for membrane current has been expanded by 10^3.

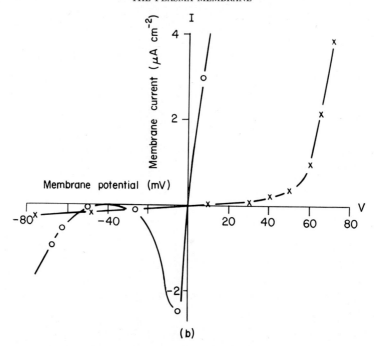

(b)

FIG. 37 (*b*). *L-V* curve for a bilayer treated with alamethicin and protamine (a poly-cation) on one side (○———○). The salt concentration (NaCl 0·1 M) is the same on both sides. The *I-V* curve of Fig. 35 has been included for comparison (×———×), but note that the scale for membrane current has been expanded × 5 and that the scale for membrane volts has been slightly reduced.

Protamine converts the rectification of Fig. 35 into a negative resistance characteristic, which is shifted with respect to the negative resistance produced by an asymmetrical distribution of NaCl (Fig. 37 (*a*)), so that there is no resting potential about the zero of membrane current. The membrane is in its conducting mode at zero applied potential, and becomes resistive as a depolarising potential of − 40 mV is applied. Reproduced by permission of Mueller and Rudin (1968), *Nature*, **217**, 713.

The kinetics of the conductance changes can be controlled by regulating the composition of the bilayer membrane;[96] thus, in a highly fluid membrane composed of unsaturated phospholipids, fast responses in the sub-millisecond range similar to the sodium and potassium systems of nerve and muscle can be achieved. In membranes formed from fully saturated phospholipids and cholesterol, rise times can be extended to as long as a second (see Fig. 40).

Single Channel Conductance Events
The approach of measuring low-level conductance fluctuations in the presence of minimal amounts of alamethicin (10⁻⁸) has been ex-plored.[94, 95, 97] In a similar manner to gramicidin A, a series of conductance

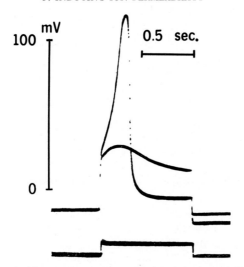

FIG. 38. Sub-threshold response and a single action potential resulting from applied depolarising pulses. The bilayer was treated with alamethicin, protamine sulphate and KCl (0·04 M) all on one side. The action potential is largely resistive in type, the resistance increasing during the spike. Reproduced by permission of Mueller and Rudin (1968), *Nature*, **217**, 713.

levels is apparent (Fig. 41), and as before the magnitude of these is quite independent of the membrane thickness, which suggests that the form of the conducting pathway depends solely on the configuration of the alamethicin aggregate. The current fluctuations of alamethicin are much greater and more rapid than those of gramicidin, and tend to occur in bursts, which commence and die away through the lowest conductance level.[94] A burst of conductivity is followed by a quiescent period. Unlike

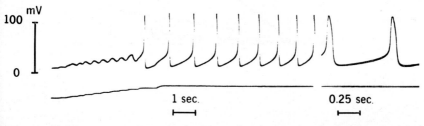

FIG. 39. Oscillating potentials developing as the current is raised above a critical level. The membrane has been treated with alamethicin, protamine sulphate and KCl (0·04 M) all on one side, and so it is in the conducting mode in the absence of current flow (see Fig. 37(*b*)). Because of the low resistance, there is only a small tendency for the membrane potential to increase as the applied current is raised initially. When the current exceeds a certain critical level, the membrane resistance rises steeply, so the potential increases, but this drives the membrane into the conducting configuration again, so the potential drops and the cycle of events repeats itself as long as the applied current is maintained. (From ref. 91.)

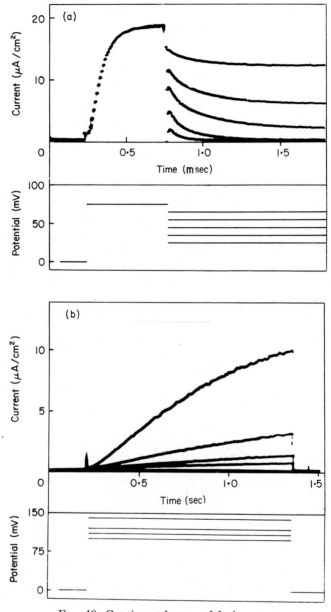

FIG. 40. *Caption at bottom of facing page.*

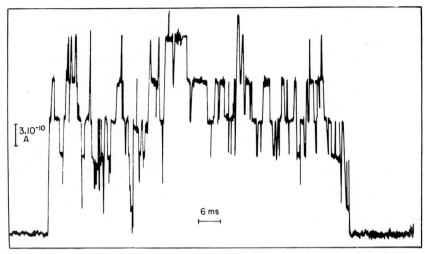

3.10^{-10} A

6 ms

FIG. 41. A single burst of current fluctuations across a bilayer membrane treated with a very small amount of alamethicin. The aqueous phase was KCl (2 M) and the applied potential was 210 mV. The baseline before and after the burst corresponds to the current flow across the unmodified membrane. Reproduced by permission of Gordon and Haydon (1972), *BBA*, **223**, 1014.

gramicidin,[36] the different conductance levels of alamethicin are not simple multiples of a single unit event[97]: by summing the conductance events of a number of bursts on a storage oscilloscope, it becomes apparent that the intermediate (3rd and 4th) levels are the most frequently occupied (see Fig. 44 (*a*)).

With one exception (alamethicin is able to sequester monovalent cations into organic solvents) the properties of alamethicin are not those of the mobile carrier ionophores and it is probable that a pore is formed. The channel conductivity based on a pore length of 30 Å (the thickness of the resistive barrier of a bilayer) suggests a diffusion coefficient for ions greater

FIG. 40. Membrane currents (above) in response to fixed potential steps (below) for two different membranes in the presence of alamethicin. Several traces have been superimposed and in (*a*) the step-down potential has been varied to demonstrate variation in the rate of decay in response to potential decrease, while in (*b*) the triggering potential has been varied.

In (*a*) the membrane was composed of a mixture of glycerol diolein (C 18:1) and diC 18:1 PO_4^-. In this unsaturated system, which is fluid at room temperature, the time constants are as low as 200 μsec, i.e. as short as the Na^+ system of nerve.

In (*b*) the membrane was composed of oxidised cholesterol and C 12:0 PO_4^{2-}. The rise times of the currents in this viscous membrane are at least 10^3 times slower than in the fluid membrane in (*a*) (note the time scales in msec and sec in (*a*) and (*b*)). The general shape of the rising phase of the current is similar, showing a brief lag and sigmoidal response in both cases. Reproduced with permission from Baumann and Mueller, *J. Supramolecular Structure*, **2**, 538.

than $1.5 \cdot 10^{-5}$ cm^2 sec^{-1},[97] which is quite comparable with the diffusion rates of K$^+$ and Cl$^-$ in aqueous solution. The pore is probably short and wide, and contains water.

The relative simplicity of the excitability inducing antibiotics, alamethicin and DJ400B, combined with the ability to regulate the details of the bilayer structures in which they are active (surface potential, viscosity, thickness, etc.) raises the possibility of understanding with some precision, the molecular basis of their activity. Quite obviously, anything which can be discussed about the mode of action of antibiotic induced excitability will be of central importance in attempting to comprehend the intrinsic excitability of muscle and nerve. At the present time, in the absence of the materials which confer excitability on natural membranes, little can be said about the molecular structures and mechanisms involved. But the close identity, in almost every detail, of the kinetic and steady-state responses of both intrinsic and antibiotic induced systems truly begs the question of whether there is a common molecular mechanism underlying all excitability phenomena.

Insertion and Aggregation

Molecular theories for electrical excitability, based mainly on the structure and properties of alamethicin, have recently been presented. The scheme discussed here is that of P. Mueller and his associates.[90, 96, 98] It appears to explain satisfactorily the main observations cited above, on alamethicin treated bilayers, and it will be seen that it follows on as an extension of the ideas developed for the polyene antibiotics, nystatin and amphotericin B. On the other hand it must be pointed out that not only are the theoretical aspects of this scheme unacceptable to some authorities[95], but even the experimental data appear to be contentious.[94] As an example of the scale of disagreement, one can point to the observation of the common exponential factor relating induced conductance to alamethicin concentration, ionic strength and membrane potential which was remarked on earlier.[91, 93] In other accounts we find that the conductance of a treated bilayer varies with the fourth power of the salt concentration but with the ninth power of the alamethicin concentration[94]; others remark on the insensitivity of the conductance to the valency of the metal ions present in the aqueous solution.[95] As mentioned above, some of these discrepancies doubtlessly arise from differences in the experimental techniques used; in particular, the functioning of alamethicin is sensitive to the composition of the lipid bilayer. We have already seen how the selectivity of the conducting channel changes from cationic to anionic when the membrane is treated with polycationic substances such as protamine, but in addition to surface charge effects it is necessary to take into consideration other factors due to the viscosity of the hydrocarbon region of the membrane, and the dimensions of the resistive layer. The scheme discussed here, then, is far from being universally accepted, and an alternative formulation for the mechanism of alamethicin action is to be found in the work of Gordon and Haydon.[93]

Mueller's theory for electrical excitability, which rests mainly on data gained from experiments with alamethicin and DJ400B (for which the structures are known) is based essentially on two events, which are termed insertion and aggregation (see Fig. 42). In the resting state it is proposed that monomeric units lie flat on the membrane surface, held there by interactions with the polar head groups of the phospholipid bilayer. The applied field, which is required to convert the system into the conducting configuration, drives them into the membrane so that they now lie in parallel with the long axes of the lipid hydrocarbon chains. This situation, which is energetically unfavourable owing to the presence of polar groups driven into the apolar region of the membrane, can become stabilised as a result of lateral diffusion of the monomers to form aggregates. By aggregating, the hydrophobic side of the antibiotic is presented outwards to the membrane hydrocarbon, while the polar aspects come together to form the conducting channel as we have seen for the polyenes, nystatin and amphotericin B. Even in its conducting aggregated state, the membrane potential is required as a stabilising force, and as soon as this is relaxed the aggregates disperse and the monomers retract to lie at the membrane-water interface. This occurs because (in contrast possibly with gramicidin and the polyenes) the pores are not stabilised by a head to head interaction with other active aggregates directed from the distal surface of the membrane, nor by interaction with the polar surface of the distal face itself. Nor is there a stabilising influence of cholesterol lying prealigned in parallel with the hydrocarbon chains of the bilayer. The presence of polar groups at the distal end of the antibiotic molecule projected right into the hydrocarbon layer can only be tolerated in the presence of the field: as soon as this is removed, the aggregates collapse.

The theory presents some limitations on the structure of putative excitability inducing molecules, and we can now examine some of the features of alamethicin and DJ400B which suit them for such a role.

Conformation of Alamethicin and DJ400B: Amphipathic Dipoles
As with the previously discussed polyene antibiotics, these molecules are probably active in an extended loop configuration — in this way their length approaches the dimensions of the hydrocarbon layer of typical membranes. With DJ400B we can see clearly the two-sided nature of the amphipathic loop, with the inflexible hydrophobic aspect on one side, and an array of hydroxyl and carbonyl oxygens on the other. The positively charged anilino group at the distal end of the molecule constitutes the gating charge, and together with the carboxyl at the proximal end, it confers on DJ400B the considerable dipole moment which is required to enable the applied field to drive the molecule across the membrane.

(a)

(b)

(c)

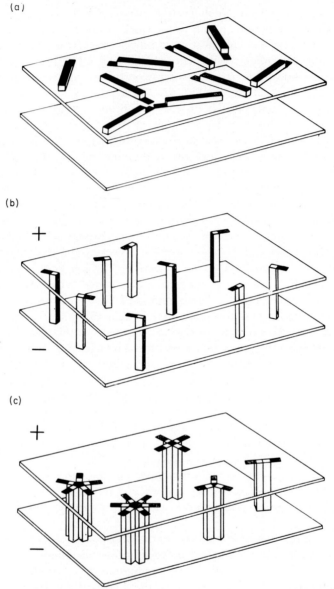

Fig. 42. An abstract model for the insertion-aggregation mechanism of the excitation process.
(a) At rest, elongated alamethicin loops lie flat on the membrane surface, which is represented by the upper plane. (b) An applied field, acting on a metal ion fixed at the distal end, has driven the alamethicin into the membrane and towards the opposite surface. (c) Lateral diffusion of monomers within the membrane leads to aggregation. Oligomeric aggregates composed of three or more alamethicin monomers form a central opening which acts as a channel for the flow of ions. Reproduced with permission from Baumann and Mueller (1975), *J. Supramolecular Structure*, **2**, 538.

In the case of alamethicin, the two sided amphipathic and dipolar quality of the molecule is not apparent from a cursory glance at the primary structure, but it is likely that the same rules apply. Luckily, the presence of seven methylalanine residues (a non-protein amino acid found in a number of peptide antibiotics) severely limits the configurational possibilities. Methylalanine

$$CH_3 \quad NH_2$$
$$\diagdown \diagup$$
$$C$$
$$\diagup \diagdown$$
$$CH_3 \quad COOH$$

(also called α-aminoisobutyric acid, AIB) is thought to be an obligatory α-helix former[99] in just the same way as certain other amino acids are helix breakers: any other configuration is stereochemically and energetically unlikely. The proposed three dimensional structure of alamethicin (see Fig. 43) is based primarily on functional considerations but it is broadly supported by spectral[100] (^{13}C-NMR and circular dichroism) and theoretical[101] (conformational energy mapping) studies. Two α-helical turns of the chain present a hydrophobic surface which could interact with the hydrocarbon interior of the membrane; the hydrophilic aspect required for the development of a channel is formed from a more flexible and extended region on the other side of the loop. The hairpin bend formed at the distal end of the elongated loop, together with the glutamine (residue 6) form a cavity lined with carbonyl oxygens which hold a metal ion in four- or five-fold coordination. This coordinated cation at the distal end of the molecule constitutes the gating charge, and together with the negatively charged glutamate residue at the proximal end, it confers on alamethicin the dipole moment, which in a similar manner to the intrinsic dipole of DJ400B, is required for the applied field to drive the molecule into the membrane.

It is this dipolar quality which initially sets these substances apart from the electrically insensitive polyene antibiotics discussed earlier. Whereas with the polyenes the energy for insertion is provided by the affinity of the antibiotics for the already aligned cholesterol, in the present case, it is the applied field which drives the positively charged centre into the hydrocarbon, aligns the molecule perpendicular to the plane of the membrane, and gives rise to a gating current. The electrophoretic movement of the antibiotic molecules into the membrane is fast.

The next step is the aggregation of the inserted monomers to form ion conducting pores. All the excitability inducing antibiotics, including alamethicin,[102] have a strong tendency towards self-aggregation, both in aqueous solution and in hydrocarbon solvents. This feature surely derives (in the case of alamethicin and DJ400B) from the very extended interface

between the polar and apolar regions of the molecules. With the phospholipids it is legitimate to talk of amphipathy arising from the different
quality of the two *ends* of an extended structure; in the present case, we are

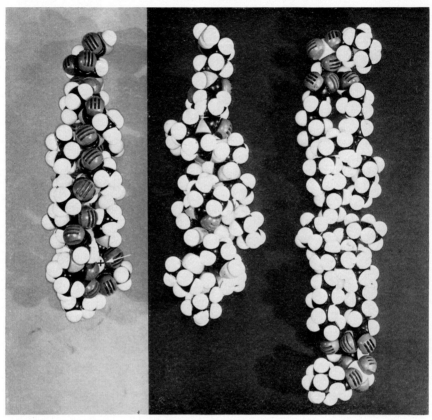

Fig. 43. Space-filling models of alamethicin, based on the cyclic structure of
Payne, Jakes and Hartley. At the right-hand side is illustrated a model of two
phosphatidyl choline molecules in bilayer configuration: it can be seen that the
loop of the alamethicin molecule approaches, but does not quite span the full
thickness of the bilayer. The fixed metal ion which constitutes the gating charge of
alamethicin is visible at the lower end of the left hand model (arrow). Reproduced
with permission of Baumann and Mueller (1975), *J. Supramolecular Structure*, **2**,
538.

concerned with the apposition of two extended *sides*, each seeking a
different environment. It is the process of aggregation which is generally
rate limiting and from which the sigmoidicity in the generation of current
flow (Fig. 40) is derived. The rate of aggregation is subject to control by
regulation of the fluidity of the membrane hydrocarbon, in a similar way
to the aggregation of the polyene antibiotics.

The length of the dipolar section of the molecule is of critical importance for the reversibility of the excitation phenomenon. On the one hand the ion channels must traverse the whole of the resistive barrier of the membrane, but on the other hand the gating charge should not come into contact with the polar surface layer and the aqueous phase at the distal side of the membrane: if this should happen, the pore could become locked in the open configuration by electrostatic forces in this region, and might not be able to relax into the non-conducting state on removal of the field. It is therefore significant that the hydrophobic portion of alamethicin in the proposed extended loop configuration, while long enough to form a competent pathway through the membrane, still falls short by about 5 Å of the lipid hydrocarbon region and must require the presence of the field to maintain the gating charge in this position. With DJ400B the position is somewhat different: the positively charged anilino group could extend right through to the distal face of the membrane, so stabilising the conducting configuration, and significantly it has been found that the closing rates of the ion conducting pathways are slow, some failing to close at all (it is also possible that the DJ400B pores could be stabilised by interaction with cholesterol, in a similar manner to nystatin and amphotericin B).

The essential operations underlying excitability induced by these substances thus appear to be:

(1) insertion of an amphipathic dipole by application of a field,
(2) aggregation of inserted dipoles to form conducting pores,
(3) disaggregation and withdrawal of monomers to the proximal surface on removal of the field.

It is on this basis that we can understand the characteristic current fluctuations which occur in the presence of alamethicin. The minimal conducting channel must be built up from at least three inserted monomers: successive recruitment of additional monomers into the minimal conducting (trimeric) aggregate will result in an expansion of the channel and a stepwise increase in conductivity. This is why the conductance events are not integral multiples of a single unit event, and are non-independent and non-random in time (cf. gramicidin A). Baumann and Mueller[94] have shown how the conductance levels of alamethicin can be simply predicted from geometrical consideration of the molecular dimensions of a prismatic channel. It is presumed that the length, l, of the conducting channel remains constant during the process of successive recruitment so that changes in conductance arise solely from changes in the area, A, which is the area of a regular polygon having n sides of side length s.

The area of the channel opening is then $A = 0.25\ ns^2 \cot(180°/n)$ and the conductivity of the n-sided channel, $G_n = \dfrac{\chi}{l}\ A$. χ is the specific

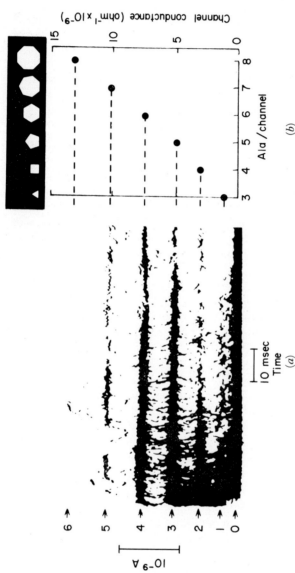

FIG. 44. (a) Current fluctuations across a bilayer membrane treated with a very small amount of alamethicin (10^{-8} M). The conditions are the same as in Fig. 41. The current pulses of many individual trains of activity have been superimposed in this figure by repetitive sweeps on a storage oscilloscope, each sweep of the trace being triggered by the leading edge of a current transition after a quiescent period. The baseline corresponds to the conductance of an unmodified bilayer. The bursts commence and terminate through the lowest levels and this accounts for the density of the superimposed traces in the lower left-hand corner of the image. Six different current levels can be distinguished and their relative probability of occurance can be estimated by the intensity of the traces at these levels: the third and fourth are the most probable. Reproduced by permission of Gordon and Haydon (1972), BBA, 223, 1014.

(b) Correlation between the conductance levels in (a) and the conductance of hypothetical prismatic channels formed by the aggregation of 3–8 alamethicin molecules. The lowest level (1) is assumed to correspond to the conductance of a trimeric aggregate, and the points represent the solution of the equation for integral values of n between 3 and 8, with the other constants adjusted to suit the conditions of the experiment in (a) as described in the text. (From ref. 96.)

conductance of the open channel ($= 0{\cdot}11\ \Omega^{-1}\ \mathrm{cm}^{-1}$ in the presence of KCl (2 M), the conditions of Gordon and Haydon's original experiment).[97] By putting $l = 25$ Å (the length of the hydrophobic portion of alamethicin) and $s = 7{\cdot}9$ Å (derived from the width of the alamethicin monomer) a quite striking parallel between predicted and measured conductance levels is obtained (Fig. 44).

This certainly lends credence to the general idea of the insertion-aggregation mechanism for alamethicin though, as has been pointed out, the distribution among a number of different conductance levels is not a direct consequence of the theory, nor is it invariably observed. It has already been remarked that the third and fourth (pentameric and hexameric) conducting levels are the most favoured. Altered conditions (e.g. different phospholipid composition) can easily alter the relative stability among the different polymeric aggregates so that only a single conductance level is apparent.

There are of course variations on this theme, and anyone familiar with the electrophysiology of muscle and nerve will be aware that whereas these two antibiotics require a potential to be driven into a conducting configuration, in cell membranes the reverse is generally the case: the action potential is initiated by the process of depolarisation. We have already shown how it is possible to shift the current-voltage curve for alamethicin along the voltage axis, by treating the bilayer membrane on the proximal side with polycationic substances such as protamine. This shift is probably due to the electrostatic repulsion between the positive charges of the polycation and the gating charge cation coordinated at the distal end of the alamethicin molecule. In this condition the ion conducting pores are open when there is no applied potential, and change to the closed condition when the field is applied. In this way the excitability induced by alamethicin resembles more closely the condition of cell membranes and bilayers which have been treated with EIM. As has been pointed out, EIM is a much more complex structure than either DJ400B or alamethicin. It has an extremely strong tendency to self association, leading to aggregate weights exceeding 10^6 in hydrocarbon solvents. In this case the ion conducting pathway is probably stabilised in the open configuration in the absence of an applied field due to the very strong affinity of EIM for itself within the hydrocarbon environment of the membrane. The membrane actually has to be polarised in the reverse direction, the gating charge being pulled out of the membrane in order to destabilise the conducting aggregate.

There is of course no strong direct evidence for an insertion-aggregation mechanism of conductance regulation in excitable biological membranes, and theories based on the allosteric regulation of enzyme activity enjoy some popularity. The assumptions involved in formulating a theoretical description of molecular events in biological membranes, in particular the sodium and potassium systems of nerve and muscle, are especially hazardous, but in the case of bilayers treated with appropriate antibiotics, the situation is much simpler. Even so, Mueller's theory for alamethicin induced excitability is by no means universally accepted (in particular the idea of accretion of alamethicin monomers into conducting channels, in order to account for multiple conductance levels) and the relevance of extending the mechanism of action of alamethicin to the problems of intrinsic excitability in biological membranes has been questioned. Going beyond the realm of electrophysiology, however, we can find evidence for aggregative processes at the cell surface controlling certain aspects of cellular activity and an example of such a process, which leads to the creation of a pathway for calcium ions, is discussed in Chapter 4.

REFERENCES

1. *Pressman, B. C. (1968). Ionophorous antibiotics as models of biological transport. *Fed. Proc.* **27**, 1283.
2. Harris, E. J. and Pressman, B. C. (1967). Obligate cation exchanges in red cells. *Nature*, **216**, 918.
3. Hopfer, U., Lehninger, A. L. and Thompson, T. E. (1968). Protonic conductance across phospholipid bilayer membranes induced by uncoupling agents for oxidative phosphorylation. *Proc. Nat. Acad. Sci.* **59**, 484.
4. Liberman, Ye. A. and Topaly, V. P. (1968). Transfer of ions across bimolecular phospholipid membranes and classification of uncouplers of oxidative phosphorylation, *Biophysics*, **13**, 1195.
5. Henderson, P. J. F., McGivan, J. D. and Chappell, J. B. (1969). The action of certain antibiotics on mitochondrial, erythrocyte and artificial phospholipid membranes. *Biochem. J.* **111**, 521.
6. Mueller, P. and Rudin, D. O. (1967). Development of K^+-Na^+ discrimination in experimental bimolecular lipid membranes by macrocyclic antibiotics. *Biochem. Biophys. Res. Commun.* **26**, 398.
7. Andreoli, T. E., Tieffenberg, M. and Tosteson, D. C. (1967). The effect of valinomycin on potassium and sodium permeability of HK and LK sheep red cells. *J. Gen. Physiol.* **50**, 2527.
8. Hope, A. B. (1971). "Ion Transport and Membranes." Butterworth, London.
9. Pressman, B. C., Harris, E. J., Jagger, W. S. and Johnson, J. H. (1967). Antibiotic mediated transport of alkali ions across lipid barriers. *Proc. Nat. Acad. Sci.* **58**, 1949.
10. Shemyakin, M. M., Vinogradova, E. I., Feignia, M. Y., Aldanova, N. A., Loginova, N. F., Ryabovia, I. D. and Pavlenko, I. A. (1965). The structure-antimicrobial relation for valinomycin depsipeptides. *Experientia*, **21**, 548.
11. Kilbourn, B. T., Dunitz, J. D., Pioda, L. A. R. and Simon, W. (1967). Structure of the K^+ complex with nonactin, a macrolide antibiotic possessing highly specific K^+ transport properties. *J. Mol. Biol.* **30**, 559.

* References marked with an asterisk (*) are mainly review articles especially recommended for further reading.

12. Ivanov, V. T., Laine, I. A., Abdolaev, N. D., Senyavina, L. B., Popov, E. M., Ovchinnikov, Yu. A. and Shemyakin, M. M. (1969). The physico-chemical basis of the functioning of biological membranes: the conformation of valinomycin and its K^+ complex in solution. *Biochem. Biophys. Res. Commun.* **34**, 803.

13. Dobler, M. (1972). The crystal structure of nonactin. *Helv. Chim. Acta,* **55**, 1371.

14. Winkler, R. (1972). Kinetics and mechanism of alkali ion complex formation in solution. *Structure and Bonding,* **10**, 1.

15. Kaplan, J. H. and Passow, H. (1974). Effects of phloridzin on net chloride movements across the valinomycin-treated erythrocyte membrane. *J. Membrane Biol.* **19**, 179.

16. Steinrauf, L. K., Pinkerton, M. and Chamberlin, J. W. (1968). The structure of nigericin. *Biochim. Biophys. Res. Commun.* **33**, 29.

17. Lutz, W. K., Winkler, F. K. and Dunitz, J. D. (1971). Crystal structure of the antibiotic monensin: similarities and differences between free acid and metal complex. *Helv. Chim. Acta,* **54**, 1103.

18. Agtarap, A., Chamberlin, J. W., Pinkerton, M. and Steinrauf, L. K. (1967). The structure of monensic acid, a new biologically active compound. *J. Amer. Chem. Soc.* **89**, 5737.

19. Pinkerton, M. and Steinrauf, L. K. (1970). Molecular structure of mono-valent metal cation complexes of monensin. *J. Mol. Biol.* **49**, 533.

20. Johnson, S. M., Herrin, J., Liu, S. J. and Paul, I. C. (1970). The crystal and molecular structure of the barium salt of an antibiotic containing a high proportion of oxygen. *J. Amer. Chem. Soc.* **92**, 4428.

21. Maier, C. A. and Paul, I. C. (1971). X-ray crystal structure of a silver complex of antibiotic X-537A; a structure enclosing two metal ions. *Chem. Comm.* 181.

22. Bissell, N. C. and Paul, I. C. (1972). Crystal and molecular structure of a derivative of the free acid of the antibiotic X-537A. *Chem. Comm.* 967.

23. *Pressman, B. C. (1973). Properties of ionophores with broad range cation specificity. *Fed. Proc.* **32**, 1698.

24. Reed, P. W. and Lardy, H. A. (1972). A23187: a divalent cation ionophore. *J. Biol. Chem.* **247**, 6970.

25. Young, S., Baker, E., Gomperts, B. D. and Huehns, E. R. (1975). Ionophore mediated iron transfer across membranes. *In* "Proteins of Iron Storage and Transport in Biochemistry and Medicine", R. R. Crighton (ed). North-Holland Publishing Co., Amsterdam.

26. Chaney, M. O., Demarco, P. V., Jones, N. D. and Occolowitz, J. L. (1974). The structure of A23187, a divalent cation ionophore. *J. Amer. Chem. Soc.* **96**, 1932.

26a. Chaney, M. O., Jones, N. D. and Debono, M. (1976). The structure of the calcium complex of A23187, a divalent cation ionophore antibiotic. *J. Antibiotics,* **29**, 424.

27. *Haydon, D. A. and Hladky, S. B. (1972). Ion transport across thin lipid membranes; a critical discussion of mechanisms in selected systems. *Quart. Rev. Biophys.* **5**, 187.

28. Chappell, J. B. and Haarhoff, K. N. (1967). The penetration of the mito-chondrial membrane by anions and cations. *In* "Biochemistry of Mito-chondria", E. C. Slater, Z. Kaninger and L. Wojtczak (eds). Academic Press, London and New York.

29. Tosteson, D. C., Andreoli, T. E., Tieffenberg, M. and Cook, P. (1968). The effects of macrocyclic compounds on cation transport in sheep red

blood cells and thin and thick lipid membranes. *J. Gen. Physiol.* **51**, 373 suppl.

30. Pressman, B. C. (1965). Induced active transport of ions into mitochondria. *Proc. Nat. Acad. Sci.* **53**, 1076.
31. Chappell, J. B. and Crofts, A. R. (1965). Gramicidin and ion transport in isolated liver mitochondria. *Biochem. J.* **95**, 393.
32. Silman, H. I. and Karlin, A. (1968). Action of antibiotics affecting membrane permeability on the electroplax. *Proc. Nat. Acad. Sci.* **61**, 674.
33. Goodall, M. C. (1970). Structural effects in the action of antibiotics on the ion permeability of lipid bilayers: gramicidins A and S, and lipid specificity. *Biochim. Biophys. Acta*, **219**. 471.
34. Hladky, S. B., Gordon, L. G. M. and Haydon, D. A. (1974). Molecular mechanisms of ion transport in lipid membranes. *Ann. Rev. Phys. Chem.* **25**, 11.
35. Myers, V. B. and Haydon, D. A. (1972). Ion transfer across lipid membranes in the presence of gramicidin A: the ion selectivity. *Biochim. Biophys. Acta*, **274**, 313.
36. Hladky, S. B. and Haydon, D. A. (1972). Ion transfer across lipid membranes in the presence of gramicidin A: studies of the unit conductance channel. *Biochim. Biophys. Acta*, **274**, 294.
37. Krasne, S., Eisenman, G. and Szabo, G. (1971). Freezing and melting of lipid bilayers and the mode of action of nonactin, valinomycin and gramicidin. *Science*, **174**, 412.
38. Sarges, R. and Witkop, B. (1965). Gramicidin A: the structure of valine and isoleucine-gramicidin A. *J. Amer. Chem. Soc.* **87**, 2011.
39. Sarges, R. and Witkop, B. (1965). Gramicidin: the structure of valine- and isoleucine-gramicidin C. *Biochemistry*, **4**, 2491.
40. Isbell, B. E., Rice-Evans, C. and Beaven, G. H. (1972). The conformation of gramicidin A in solution. *FEBS Lett.* **25**, 192.
41. Veatch, W. R., Fossel, E. T. and Blout, E. R. (1974). The conformation of gramicidin A. *Biochemistry*, **13**, 5249.
42. Urry, D. W. (1972). Protein conformation in biomembranes: optical rotation and absorption of membrane suspensions. *Biochim. Biophys. Acta*, **265**, 113.
43. Urry, D. W. (1971). The gramicidin A transmembrane channel: a proposed π_{LD} helix. *Proc. Nat. Acad. Sci.* **68**, 672.
44. Urry, D. W., Goodall, M. C., Glickson, J. D. and Mayers, D. F. (1971). The gramicidin A transmembrane channel; characteristics of head-to-head dimerized $\pi_{(L.D.)}$ helices. *Proc. Nat. Acad. Sci.* **68**, 1907.
45. *Kinsky, S. C. (1970). Antibiotic interaction with model membranes. *Ann. Rev. Pharmacol.* **10**, 119.
46. Borowski, E., Zielinsky, J., Falkowski, L., Ziminski, T., Golik, J., Kotodziejczyk, P., Jereczek, E. and Golulewicz, M. (1971). The complete structure of the polyene macrolide antibiotic nystatin A_1. *Tetrahedron Lett.* 685.
47. Bergy, M. E. and Eble, T. E. (1968). The filipin complex. *Biochemistry*, **7**, 653.
48. Borowski, E., Zielinski, J., Ziminski, T., Falkowski, L., Kotodziejczyk, P., Golik, J. and Jereczek, E. (1970). Chemical studies with amphotericin B: the complete structure of the antibiotic. *Tetrahedron Lett.* 3909.
49. Kinsky, S. C. (1963). Comparative responses of mammalian erythrocytes and microbial protoplasts to polyene antibiotics and vitamin A. *Arch. Biochem. Biophys.* **102**, 180.

50. Lampen, J. O. (1966). Interference by polyenic antifungal antibiotics (especially nystatin and filipin) with specific membrane functions. *Symp. Soc. Gen. Microbiol.* **16**, 111.

51. Cass, A. and Dalmark, M. (1973). Equilibrium dialysis of ions in nystatin-treated red cells. *Nature*, **244**, 47.

52. Kinsky, S. C., Luse, S. A. and van Deenen, L. L. M. (1966). Interaction of polyene antibiotics with natural and artificial membrane systems. *Fed. Proc.* **25**, 1503.

53. Verkleij, A. J., de Kruijff, B., Gerritsen, W. F., van Deenen, L. L. M. and Ververgaert, P. H. J. (1973). Freeze-etch electron microscopy of erythrocytes, *Acholeplasma laidlawii* cells and liposomal membranes after the action of filipin and amphotericin B. *Biochim. Biophys. Acta*, **291**, 577.

54. Kinsky, S. C., Luse, S. A., Zopf, D., van Deenen, L. L. M. and Haxby, J. (1967). Interaction of filipin and derivatives with erythrocyte membranes and lipid dispersions: electron microscopic observations. *Biochim. Biophys. Acta*, **135**, 844.

55. Demel, R. A., van Deenen, L. L. M. and Kinsky, S. C. (1965). Penetration of lipid monolayers by polyene antibiotics: correlation with selective toxicity and mode of action. *J. Biol. Chem.* **240**, 2749.

56. Demel, R. A., Crombag, F. J. L., van Deenen, L. L. M. and Kinsky, S. C. (1968). Interaction of polyene antibiotics with single and mixed monomolecular films. *Biochim. Biophys. Acta*, **150**, 1.

57. Lampen, J. O., Arnow, P. M., Borowska, Z. and Laskin, A. I. (1962). Location and role of sterol at nystatin binding sites. *J. Bact.* **84**, 1152.

58. de Kruijff, B., Gerritsen, W. J., Oerlemans, A., Demel, R. A. and van Deenen, L. L. M. (1974). Polyene antibiotic sterol-interactions in membranes of *Acholeplasma laidlawii* cells and lecithin liposomes: specificity of the membrane permeability changes induced by the polyene antibiotics. *Biochim. Biophys. Acta*, **339**, 30.

59. van Zutphen, H., Demel, R. A., Norman, A. W. and van Deenen, L. L. M. (1971). The action of polyene antibiotics on lipid bilayer membranes in the presence of several cations and anions. *Biochim. Biophys. Acta*, **241**, 310.

60. Andreoli, T. E. and Monahan, M. (1968). The interaction of polyene antibiotics with thin lipid membranes. *J. Gen. Physiol.* **52**, 300.

61. Cass, A., Finkelstein, A. and Krespi, V. (1970). The ion permeability induced in thin lipid membranes by the polyene antibiotics nystatin and amphotericin B. *J. Gen. Physiol.* **56**, 100.

62. Dennis, V. W., Stead, N. W. and Andreoli, T. E. (1970). Molecular aspects of polyene- and sterol-dependent pore formation in thin lipid membranes. *J. Gen. Physiol.* **55**, 375.

63. Holz, R. and Finkelstein, A. (1970). The water and nonelectrolyte permeability induced in thin lipid membranes by the polyene antibiotics nystatin and amphotericin B. *J. Gen. Physiol.* **56**, 125.

64. Andreoli, T. E., Dennis, V. W. and Weigl, A. M. (1969). The effect of amphotericin B on the water and nonelectrolyte permeability of thin lipid films. *J. Gen. Physiol.* **53**, 133.

65. *Finkelstein, A. and Cass, A. (1968). Permeability and electrical properties of thin lipid membranes. *J. Gen. Physiol.* **52**, 145, supplement.

66. van Zutphen, H., van Deenen, LL.M. and Kinsky, S. C. (1966). The action of polyene antibiotics on bilayer lipid membranes. *Biochem. Biophys. Res. Commun.* **22**, 393.

67. Marty, A. and Finkelstein, A. (1975). Pores formed in lipid bilayer membranes by nystatin. *J. Gen. Physiol.* **65**, 515.

68. *Cass, A. and Finkelstein, A. (1967). Water permeability of thin lipid membranes. *J. Gen. Physiol.* **50**, 1765.
69. Razin, S., Tourtellotte, M. E., McElhaney, R. N. and Pollack, J. D. (1966). Influence of lipid components of *Mycoplasma laidlawii* on osmotic fragility of cells. *J. Bact.* **91**, 609.
70. de Kruijff, B., Demel, R. A. and van Deenen, L. L. M. (1972). The effect of cholesterol and epicholesterol incorporation on the permeability and the phase transition of intact *Acholeplasma laidlawii* cell membranes and derived liposomes. *Biochim. Biophys. Acta*, **255**, 331.
71. Demel, R. A., Bruckdorpher, K. R. and van Deenen, L. L. M. (1972). Structural requirements of sterols for the interaction with lecithin at the air-water interface. *Biochim. Biophys. Acta*, **255**, 311.
72. Norman, A. W., Demel, R. A., de Kruijff, B. and van Deenen, L. L. M. (1972). Studies on the biological properties of polyene antibiotics: evidence for the direct interaction of filipin with cholesterol. *J. Biol. Chem.* **247**, 1918.
73. Norman, A. W., Demel, R. A., de Kruijff, B., Geurts van Kessel, W. S. M. and van Deenen, L. L. M. (1972). Studies in the biological properties of polyene antibiotics: comparison of other polyenes with filipin in their ability to interact specifically with sterol. *Biochim. Biophys. Acta*, **290**, 1.
74. de Kruijff, B., Gerritsen, W. J., Oerlemans, A., van Dijck, P. W. M., Demel, R. A. and van Deenen, L. L. M. (1974). Polyene antibiotic-sterol interactions in membranes of *Acholeplasma laidlawii* cells and lecithin liposomes: temperature dependence of the polyene antibiotic-sterol complex formation. *Biochim. Biophys. Acta*, **339**, 44.
75. *de Kruijff, B. and Demel, R. A. (1974). Polyene antibiotic-sterol interactions in membranes of *Acholeplasma laidlawii* cells and lecithin liposomes: molecular structure of the polyene antibiotic-cholesterol complexes. *Biochim. Biophys. Acta*, **339**, 57.
76. Finkelstein, A. and Holz, R. (1973). Aqueous pores created in thin lipid membranes by the polyene antibiotics nystatin and amphotericin B. *In* "Membranes, vol. 2: Lipid Bilayers and Antibiotics", G. Eisenman (ed). Marcel Dekker Inc., New York.
77. *Andreoli, T. E. (1973). On the anatomy of amphotericin B-cholesterol pores in lipid bilayer membranes. *Kidney International*, **4**, 337.
78. Finkelstein, A. (1964). Electrical excitability of isolated frog skin and toad bladder. *J. Gen. Physiol.* **47**, 545.
79. Pickard, B. G. (1974). Electrical signals in higher plants. *Naturwiss.* **61**, 60.
80. Osterhout, W. J. and Hill, S. E. (1939). Pacemakers in *Nitella*: arrhythmia and block. *J. Gen. Physiol.* **22**, 113.
81. Eckert, R. (1972). Bioelectric control of ciliary activity: locomotion in the ciliated protozoa is regulated by membrane-limited calcium fluxes. *Science*, **176**, 473.
82. Maeno, T. (1938). Electrical characteristics and activation potential of *Bufo* eggs. *J. Gen. Physiol.* **43**, 139.
83. Yamamoto, T. (1961). Physiology of fertilization in fish eggs. *Int. Rev. Cytol.* **12**, 361.
84. Mueller, P., Rudin, D. O., Tien, H. T. and Wescott (1962). Reconstitution of excitable cell membrane structure *in vitro*. *Circulation*, **26**, 1167.
85. Kushnir, L. D. (1968). Studies in a material which induces electrical excitability in bimolecular lipid membranes: production, isolation, gross identification and assay. *Biochim. Biophys. Acta*, **150**, 285.

86. Payne, J. W., Jakes, R. and Hartley, B. S. (1970). The primary structure of alamethicin. *Biochem. J.* **117**, 757.

87. Martin, D. R. and Williams, R. J. P. (1975). The nature and function of alamethicin. *Biochem. Soc. Trans.* **3**, 166.

88. Bohlmann, F., Dehmlow, E. V., Neuhahn, H. J., Brandt, R. and Bethke, H. (1970). New heptaene macrolide: primary structure, arrangement of the functional groups and structure of the aglycone (in German). *Tetrahedron*, **26**, 2199.

89. *Mueller, P. and Rudin, D. O. (1968). Translocation in bimolecular lipid membranes: their role in dissipative and conservative bioenergy transductions. *In* "Current Topics in Bioenergetics, Vol. 3", D. R. Sanadi (ed). Academic Press, New York and London.

90. *Mueller, P. (1975). Electrical excitability in lipid bilayers and cell membranes. *In* "Energy Transducing Mechanisms", E. Racker (ed). M.T.P. International Review of Science, Biochemistry series 1, vol. 3. Butterworth, London.

91. *Mueller, P. and Rudin, D. O. (1968). Action potentials induced in bimolecular lipid membranes. *Nature*, **217**, 713.

92. Cherry, R. J. (1972). Model membranes and excitability. *Chem. Phys. Lipids*, **8**, 393.

93. Cherry, R. J., Chapman, D. and Graham, D. E. (1972). Studies of the conductance changes induced in bimolecular lipid membranes by alamethicin. *J. Membrane Biol.* **7**, 323.

94. Eisenberg, M., Hall, J. E. and Mead, C. A. (1973). The nature of the voltage-dependent conductance induced by alamethicin in black lipid membranes. *J. Membrane Biol.* **14**, 143.

95. Gordon, L. G. M. and Haydon, D. A. (1975). Potential-dependent conductances in lipid membranes containing alamethicin. *Phil. Trans. R. Soc. Lond. B.* **270**, 433.

96. *Baumann, G. and Mueller, P. (1975). A molecular model of electrical excitability. *J. Supramolecular Structure*, **2**, 538.

97. Gordon, L. G. M. and Haydon, D. A. (1972). The unit conductance channel of alamethicin. *Biochim. Biophys. Acta*, **233**, 1014.

98. Mueller, P. (1975). Membrane excitation through voltage-induced aggregation of channel precursors. *Ann N.Y. Acad. Sci.* **264**, 247.

99. Burgess, A. W. and Leach, S. J. (1973). An obligatory α-helical amino acid residue. *Biopolymers*, **12**, 2399.

100. Jung, Q., Dubischar, N. and Liebfritz, D. (1975). Conformational changes of alamethicin induced by solvent and temperature. *Eur. J. Biochem.* **54**, 395.

101. Burgess, A. W. and Leach, S. J. (1973). Conformational studies on alamethicin. *Biopolymers*, **12**, 2691.

102. McMullen, A. E. and Stirrup, J. A. (1971). The aggregation of alamethicin. *Biochim. Biophys. Acta*, **241**, 807.

4 | Liposomes and Ionophores as Tools in Cell Biology

INTRODUCTION

Earlier chapters in this book have demonstrated the validity and power of model structures as tools in the investigation of the architecture and function of biological membranes. By the use of well-defined small molecules, mainly of microbiological origin, these inert structures can be modified so as to mimic very closely some of the highly developed features of the membranes of differentiated cells. It has frequently been suggested that this approach holds out the greatest promise in the attempt to understand the molecular basis of membrane function.

It is the purpose of this chapter to show, by way of examples, how model structures can be used as tools in the realm of cell biology to solve problems which would probably be inaccessible by any other means. Two examples have been selected in order to illustrate the application of both a model membrane system and the ion-transporting antibiotics, and in each case it is necessary to provide some of the general background so that it becomes clear just what kind of questions can best be addressed by the use of the models. The specific examples, the phenomenon of cell lysis mediated by antibody and complement, and the triggering of the exocytotic release of histamine from rat mast cells are both highly specialised interests in their own right. They are none the less both close to the heart of current thinking in cell biology and form part of a much wider experience than their given titles appear to convey.

CELL LYSIS BY ANTIBODY AND COMPLEMENT

General Background
When cells are treated with fresh serum containing antibodies directed

at antigenic structures on the cell sufrace, the cell membrane becomes irreversibly damaged and the cell lyses. On the other hand, if the antibody-containing serum is first heated at 56° for 30 min, no lysis occurs although it can be demonstrated that attachment of the antibody still takes place and subsequent addition of fresh serum from a source free of antibody (the "complement" serum) is once again able to induce cell lysis. Paul Erhlich[1] in 1899 defined complement as that activity of blood serum which completes the action of antibody, but we recognise now that it has many other diverse functions. Nearly all of these have their effects on cell membranes.[2] As well as being responsible for cell lysis, complement factors are able to activate specialised cell functions such as histamine secretion from mast cells and platelets,[3, 4] directed migration (chemotaxis of leucocytes,[5,6] aggregation and fusion of platelets[7] and the contraction of smooth muscle.[8] Certain diseases such as immune vasculitis and nephritis, lupus erythematosus, rheumatoid arthritis and glomerulonephritis are probably pathogenic manifestations of the activation of complement *in vivo*.[9, 10] On the other hand, a required physiological role for complement can be discerned in individuals having inherited deficiencies of complement function. Complement deficiency states have been described both in man[11, 12] and in animals[13, 14] and an increased susceptibility towards bacterial infection has been indicated in the case of complement-deficient mice. Intact complement then confers advantages in combating diseases.

Assembling the Membrane Attack Complex

The mechanism of complement lysis seems to be a perfectly general one, given a suitable antigen-antibody at the cell surface, and a source of active serum; there is no cell type which is endowed with immunity against it. A requirement for eleven separate serum glycoproteins has been demonstrated, and these react with each other in defined order as a "cascade" having some superficial resemblances to the processes of protein-protein interaction in blood coagulation; this is illustrated in the scheme below. The process will be described in only the most elementary outline; the protein chemistry of complement has been excellently reviewed by Muller-Eberhard,[2, 15] who has himself been responsible for elucidating the details of large areas of this most complex field.

All of the complement proteins are soluble in aqueous solution, and none of them has any specialised affinity for the surface of membranes. After enzyme mediated activation, three of the components of complement become capable of entering into direct and firm contact with the membrane surface, and the enzymes involved (also complement proteins) are themselves activated by binding to surface-bound activator units. The first protein of the sequence, C1q a high molecular weight protein having 18 subunits, has the role of recognising immune complexes, and in so doing it initiates the complement reaction. Two other complement (C) proteins,

C1r and C1s associate reversibly with C1q and this leads via internal reactions to the generation of the active species $\overline{C1s}$, which is directed against C4 and C2, splitting them both into two fragments, a and b. The fragments C4a and C2b are discharged into solution and C4b attaches to C2a to generate the active species $\overline{C4, 2}$ or C3-convertase which binds to the cell surface at a point removed from the original point of attachment of C1. This complex splits C3 into two fragments, and combines with C3b

---

Scheme of complement mediated cytolysis: functional units and membrane sites[15]

First site: Activation of Recognition Unit

1. site (1) $Ab + C1q\left\langle{}^{r}_{s}\right. \rightarrow$ site (1) $Ab - C1q\left\langle{}^{r\ \downarrow}_{s}\right.$

Second site: Assembly of Activation Unit

2. C4 $\xrightarrow{\text{site (1) Ab}\overline{\text{C1}}}$ C4a + C4b*

3. C2 $\xrightarrow{\text{site (1) Ab}\overline{\text{C1}}}$ C2a* + C2b

4. site (2) + C4b* + C2a* → site (2) $\overline{\text{C4b, 2a}}$

5. C3 $\xrightarrow{\text{site (2) }\overline{\text{C4b, 2a}}}$ C3a + C3b*

6. site (2) $\overline{\text{C4b, 2a}}$ + C3b* → site (2) $\overline{\text{C4b, 2a, 3b}}$

Third site: Assembly of Membrane Attack Mechanism

7. C5 $\xrightarrow{\text{site (2) }\overline{\text{C4b, 2a, 3b}}}$ C5a + C5b*

8. site (3) + C5b* + C6 + C7 + C8 + C9 → site (3) C5b, 6, 7, 8, 9

N.B. Sites 1, 2, and 3 are considered to be at topographically distinct locations on the membrane surface. Ab is the antibody to an antigenic component of the membrane surface. A bar over a numbered complement factor denotes that it is an active enzyme capable of altering (rather than just binding to) another factor in the series. An asterisk denotes an enzymically activated binding site.

---

to form the active species $\overline{C4, 2, 3}$ which constitutes the activation unit; this is active against C5. The membrane attack complex is assembled in solution following the cleavage of C5 into two fragments, a and b. C5 is only transiently active, but enters into stable association with C6 and C7. Up to this point there is no evidence that irreversible membrane damage has occurred, but following the addition of a single molecule of C8 to the C5b, 6, 7 complex, it is possible to detect a small degree of cell lysis in a sensitive system (leakage of haemoglobin from red blood cells). Addition of six molecules of C9 per C5b–8 complex greatly accelerates the rate of cell lysis. The attack complex is visualised as a trimolecular C5b, 6, 7 complex

G

resting on the cell surface, which forms the binding site for one molecule of C8, the molecule which actually executes the lytic *coup de grace*. C9 appears to enhance the expression of the lytic activity of C8.

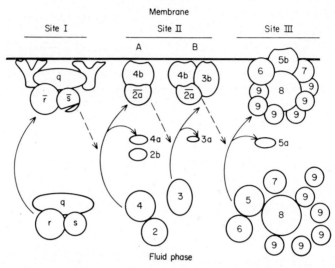

FIG. 1. Pictorial representation of the three site process of complement transfer from solution to the target cell surface. The C1 complex is reversibly bound to antibody molecules at site (I) through the C1q subunit. An internal reaction leads to activation of C1s. C1s activates labile binding sites in C2 and C4 by enzymic removal of activation peptides. The C4, 2 (C3-convertase) is thereby enabled to assemble at a membrane site (II) which is topographically distinct from site (I). C4, 2 reacting with C3, is converted to C4, 2, 3 (C5-convertase) by assimilation of C3b. C4, 2, 3 by cleaving C5 initiates the self assembly of the C5b–9 complex which attacks the membrane at site (III). Reproduced by permission of Müller-Eberhard (1975), *Ann. Rev. Biochem.* **44**, 697.

Lysis following complement damage is generally thought to be an osmotic phenomenon, which is preceded by a flow of ions, the flow of K^+ out of the cell being less than the inward flow of Na^+; water then moves into the cell causing it to swell and rupture[16, 17] (though this mechanism has been questioned).[18] The final event, the release of the cell contents (but not the equilibration of the ion gradients) can be prevented by allowing complement damage to take place in the presence of a high concentration of protein which balances the colloid osmotic pressure of the cell interior.

Electron Microscopic Lesions: Pores or Wells?
The damage to cell membranes caused by the action of complement can be observed at the ultrastructural level by means of electron microscopy. Negative stained preparations reveal the presence of numerous lesions or

"pits" about 100 Å in diameter[19, 20] (see Fig. 2). It will be recalled that in negative stain electron microscopy, the electron-dense stain accumulates in regions accessible to water, and so a lot of attention has been directed in an attempt to demonstrate that the "pits" might be the superficial expression of pores which could penetrate right across the membrane, and through which ions and small molecules could flow. The lesions in complement damaged membranes bear a rather close resemblance to those

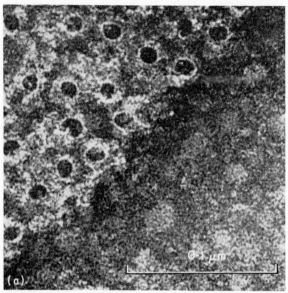

FIG. 2 (a). Negative stain electron microscope photograph of a red cell membrane treated with antibody and complement. The appearance and dimensions of the lesions vary little, whichever type of membrane, antibody or complement is used. The lesions appear as a dark central hole (diameter 85–110 Å) or indentation filled with negative stain which is surrounded by a clear ring which is raised above the plane of the membrane surface. (Original electron microscope photograph by Dr Robert Dourmashkin.)

seen in cholesterol-containing membranes and liposomes after damage by the action of the polyene antibiotic filipin (Chapter 3, p. 145) but it will be recalled that in that case the weight of evidence was certainly against the idea of a transmembrane pore. However, there is an excellent correlation between the number of lesions and the extent of membrane damage, and under conditions of partial lysis in a cell preparation it is possible to demonstate that the dark stained pits are only present on lysed cells. In agreement with the "single hit" hypothesis[21] for complement lysis (which states that a single fixed immune complex is sufficient to cause cell lysis) there is statistical evidence to suggest that the presence of but a single pit may be a sufficient lesion to cause irreversible cell damage. The technique of freeze

G*

fracture electron microscopy (see Chapter 2, pp. 68 *et seq.*) has also been applied to the investigation of complement damage.[22] This has shown that whilst it is possible to detect alterations due to complement within the

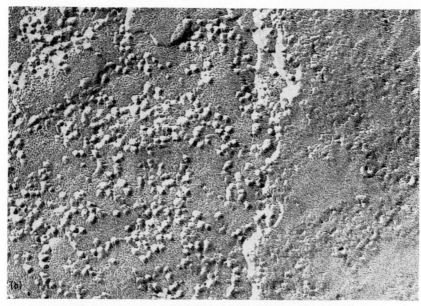

FIG. 2 (*b*). Freeze-etch appearance of sheep red cell membranes which have been lysed by a complete system of antibody and complement. The ES (extracellular) surface on the right is altered by the presence of a number of surface rings— probably the same manifestation as is revealed by negative staining procedures. The distribution of intercalated particles on the PF inner fracture face is not detectably different from the PF face of an untreated cell (see for example Fig. 12, Chapter 2). (From ref. 22.)

fracture plane, these are all limited to the extracellular EF fracture face. The lesion does not penetrate right across the bilayer, and if the structure of the lytic lesion is indeed a channel, then we have to conclude that its dimensions are below the size at present detectable by ultrastructural techniques.

LIPOSOMES: MODEL TARGETS[23]

With the ultrastructural approach running into sterile ground, the point was reached where knowledge of the protein chemistry of complement far outstripped the understanding of the mechanism of cell damage. By using cells as the substrate for studying complement action it was difficult to know which structures in the membrane are the actual targets for attack by the activated complex. In terms of the universality of complement lysis against all cell types, it would seem likely that the target would be a universal component of membrane structure, and for this reason alone the permeability barrier of the phospholipid bilayer would seem to be a plausible candidate, though a protein target cannot be excluded. It was

FIG. 2 (c). The PS inner etch face is quite smooth – there is no sign of the surface rings which characterise the outer ES face; the EF (outer fracture) face reveals signs of alteration in the form of discrete aggregates composed of three or four globules tightly fused together. The surface density of the globule aggregates (284 μm^{-2}) on the EF fracture plane correlates closely with the density of surface rings on the ES external surface (255 μm^{-2}) shown in (b). None of these alterations, which are exclusively located on the external leaflet of the membrane, were observed when the lytic system lacked either antibody or complement. Reproduced by permission of Iles et al. (1973), J. Cell. Biol. **56**, 568.

for reasons such as these, as well as the similarity of appearance of the lesions induced in membranes by complement, and in membranes and liposomes by filipin, that S. C. Kinsky and his colleagues first tried to see if they could develop immunologically responsive liposomes, that could be lysed by complement.[21]

Measurement of Liposomal "Lysis"

The first requirement was for a method for measuring the lysis of liposomal membranes. Most of the experiments have been conducted with liposomes loaded with glucose as an internal marker, and this is included in the aqueous solution used to hydrate the phospholipid. It will be recalled (Chapter 1, p. 29) that phospholipid liposomes are highly impermeable to glucose, so that after washing the liposomes (by centrifugation) the only way glucose can leak out is by damaging the permeability barrier (e.g. by treatment with filipin). The loss of glucose can then be detected by an enzyme-catalysed reaction coupled to the generation of NADPH, which is measured by fluorescence or its absorption at 340 nm.

$$\text{glucose} + \text{ATP} \xrightarrow{\quad\text{hexokinase}\quad} \text{glucose-6-phosphate} + \text{ADP}$$

$$\text{glucose-6-phosphate} + \text{NADP}^+ \xrightarrow{\quad\begin{array}{c}\text{glucose-6-phosphate}\\ \text{dehydrogenase}\end{array}\quad} \text{6-phosphogluconate} + \text{NADPH} + \text{H}^+$$

The method has a number of advantages over alternative methods involving measurement of released isotopically labelled markers, the main advantage being that the loss of internal marker may be detected in the presence of the liposomes, since the unreleased material entrapped in the liposomes is inaccessible to the enzymes used in the assay. Thus both the reaction under investigation (complement lysis) and the assay (measurement of glucose release) may be measured simultaneously by using the cuvette as the reaction vessel. A minor disadvantage of the method is that any glucose present in the serum used as a source of complement must be removed (by dialysis) before use. Even this step can be simply overcome by substituting the fluorogenic substrate β-methyl umbelliferyl phosphate for glucose as the internal marker.[25] This yields an intensely fluorescent product after hydrolysis catalysed by alkaline phosphatase which can be added to the external fluid.

Preparation of Immunologically Responsive Liposomes

(a) *Sheep Red Cell Lipids and the Forssman Antigen.* At first, crude lipid extracts were prepared from sheep red cell membranes.[24] Such extracts, which contain the phospholipid and cholesterol, also contain an antigen (the Forssman antigen) which is itself an amphipathic glycolipid against which the main antibody present in the sera of animals immunised with sheep red blood cells is directed. The hope that the Forssman antigen could be included into the bilayer structure of well-sealed liposomes prepared from a crude lipid extract in an immunologically expressible manner proved fully justified. Over a thirty-minute period, only insignificant amounts of the internally trapped glucose were lost when the liposomes were suspended in the presence of an antibody (a heat-inactivated rabbit serum containing immunoglobulin M directed against sheep red cells) or an active source of complement (fresh or carefully preserved guinea-pig serum). By contrast, when the liposomes were suspended in a medium containing both the antibody and complement sera, then about 65% of the entrapped glucose was released in a short period of time (see Fig. 3).

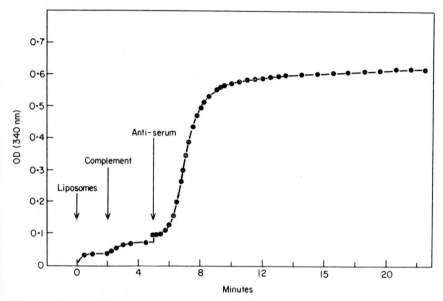

FIG. 3. Spectrophotometric assay for complement mediated lysis of antigenic liposomes, loaded with glucose. At time zero, a small quantity of liposomes prepared from sheep red cell lipids (containing the Forssman antigen and which have been loaded with glucose as an internal marker) were added to the cuvette. The solution in the cuvette contained ATP, NADP$^+$ and the enzymes hexokinase and glucose-6-phosphate dehydrogenase. The presence of free glucose initiates reactions leading to the generation of NADPH, which results in an increase in the optical density at 340 nm. The addition of complement serum after 2 min produced a very slight increment in OD, which was most probably due to residual glucose present in the serum remaining after dialysis. Addition of antiserum initiates a large increase in OD after a short lag period which is characteristic of complement mediated lysis. (From ref. 24.)

(b) *Relevance of the Method.* Several lines of evidence suggested that the process of glucose release from sheep red cell lipid liposomes closely resembles complement lysis of cells. The involvement of an antigen-antibody reaction at the liposome surface was demonstrated by the use of liposomes prepared from bovine-red cell lipids. These do not contain the Forssman antigen and in consequence the rabbit-anti-sheep antibody plus guinea-pig serum was ineffective in causing leakage of the marker glucose. Serum from C6-deficient rabbits was inactive as a source of lytic activity,[26] and a requirement for other individual proteins of complement has been demonstrated.[27] Further similarities between the complement lysis of red blood cells and the loss of glucose from liposomes have been described.[23, 24] For instance, the dependence of the extent of the reaction on the amount of antibody (a hyperbolic relationship) and on the amount of complement (a sigmoidal relationship) demonstrated a close identity of mechanism. The kinetics of glucose release from liposomes and of cell lysis both display a lag phase (Fig. 3), regardless of whether the reaction is initiated experimentally by the addition of the antiserum, the complement serum, or the liposomes (or cells) to the other components. The lag phase in both instances is shortest when the reaction is initiated by addition of complement, indicating that the antigen-antibody complex must be formed first before the complement sequence can be activated. If the complement serum is heated at 46°, its ability to cause glucose loss from liposomes or to cause cell lysis is lost in a similar time-dependent manner over 90 min. All these points of identity fully justified the hope that the model membrane system would prove an appropriate vehicle for the study of complement lysis, and the way was now open for a more detailed study of mechanism, especially in terms of controlled phospholipid and antigen composition.

(c) *Antigenic Liposomes of Defined Composition.* As in other experimental strategies involving model membranes, the lipid composition could be varied at will, and liposomes were prepared from a mixture of phospholipid (generally sphingomyelin appeared to be more suitable than phosphatidyl choline), a charged amphipathic substance (diC16 phosphate or stearylamine, $C18NH_2$) and cholesterol.[28] The nature of the antigen and its proportion to phospholipid could also be varied. Experiments were initially carried out with purified Forssman antigen whose structure[29] is shown in Fig. 4. The amphipathic nature of this substance is

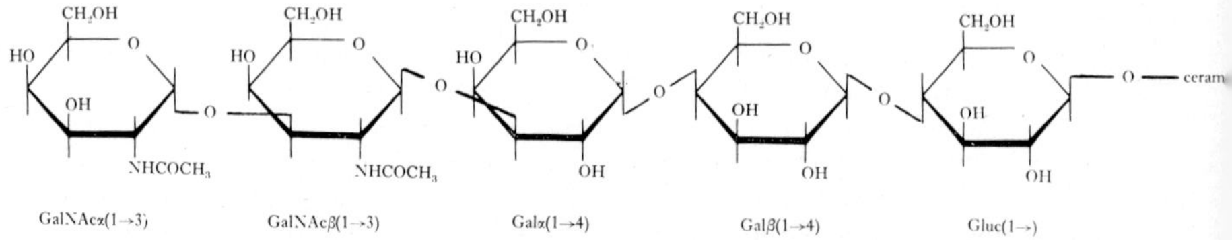

FIG. 4. Structure of the Forssman antigen. (From ref. 29.)

controlled by the two long hydrocarbon chains of the ceramide portion at one end of the molecule, and by the hydrophilic pentasaccharide which carries the antigenic determinant at the other. As with the earlier experiments using liposomes prepared from crude lipid extracts of sheep red cell membranes, the pure lipid system proved to be an entirely relevant model for the study of complement lysis.

It was found that the amphipathic antigen had to be incorporated into the liposomes at the time of their generation (active sensitisation). It was not possible to detect marker loss when the antigen was added to preformed liposomes ("passive sensitisation") indicating that the antigen is incorporated right into the bilayer structure. By varying the proportion of Forssman antigen added to the system, it was shown that a critical amount of the antigen must be incorporated into the liposomal membrane before any marker loss can be detected as a result of subsequent treatment by antibody and complement. Thereafter, as the proportion of the antigen is increased, the extent and rate of marker loss increases too, until maximum sensitisation is achieved at a ratio of about one antigen per 250 sphingo-myelin molecules[30] when antibody molecules of the IgM class are used to initiate the reaction (the ratio is different for IgG). In a converse sense, in order to achieve a given degree of marker release in systems having a high antigen concentration there is a reduced requirement for antibody, indicating that the extent of marker loss is dependent on the number of antigen-antibody complexes formed. Further evidence for the dependence of marker release on the number of immune complexes fixed at the membrane surface came from experiments in which the liposomes were prepared having another ceramide antigen in addition to the Forssman.[30] This second antigen, human globoside 1, is the predominant ceramide antigen of human red blood cells, and in spite of its very close structural relationship to the Forssman substance, individual antisera can be pro-duced which show no cross reactivity. (The specificity of Forssman anti-genicity appears to derive from the terminal $\alpha(1-3)$-glycosidic linkage between the two N-acetyl galactosyl groups: removal of the terminal residues results in antigenic and chemical properties indistinguishable from human globoside 1.)

A marker release experiment was carried out on the doubly sensitised liposomes using less than the optimal amount of anti sheep red cell (anti-Forssman) serum in order to limit the extent of marker release. Subsequent addition of anti-human red cell serum (anti-globoside) which generates a separate class of immune complexes on the liposomal surface then led to renewed release of the entrapped marker. An important point to note here is that experiments of this sort, in which the nature and proportions of the active antigens can be varied at will, would hardly be practicable in bio-logical systems. So long as we accept the relevance of the model membrane to the biological process under study, it is capable of giving us much more precise information than could ever be obtained from experiments with natural systems, but the greatest care should be taken to understand even the smallest anomalies that arise.

(d) *Further Comments on the Relevance of the Method.* The experiments cited above could be interpreted in a manner which is at variance with the generally accepted idea that a single complement lesion on a cell membrane is sufficient to cause cell lysis, since the extent of marker loss from lipo-somes is clearly regulated in a proportional manner by the number of fixed immune complexes. Certainly the loss of the small molecular weight marker from complement damaged liposomes should not be equated with the loss of haemoglobin from red cells. Lysis, the final all or none event in the sequence of events triggered by antibody and complement is an osmotic phenomenon, and there is no reason to expect that the liposomes in these experiments will be subject to extensive disruption of this sort. There is another relevant difference between cells and liposomes: that is that all of the antigenic sites of the biological surface are exposed to the exterior, whereas in liposomes of multilamellar structure, only the antigenic sites of the outermost membrane are available for reaction with antisera. It is for this reason that total release of entrapped marker glucose is never achieved although release of the order of about 65% is commonly observed from sphingomyelin liposomes and this is certainly more than could be contained in the outermost aqueous compartment of a multibilayer liposome. (In an idealised concentric liposome having 30 equispaced layers, Kinsky has estimated[23] that 10% of the total marker would be contained in the outermost compartment. The seven outer compartments together contain about 50% of the marker.)

Synthetic Hapten Antigens
Some clues to the mechanism whereby marker release from buried regions of the liposome can occur have come from investigations using synthetic (hapten) antigens[31] such as dinitrophenol (DNP). The polar group of phosphatidyl ethanolamine terminates in an uncomplicated amino group which may be reacted to form DNP phosphatidyl ethanolamine, DNP-PE

$$R-\overset{\overset{\displaystyle O}{\|}}{C}-O-CH_2$$

$$R-\underset{\underset{\displaystyle O}{\|}}{C}-O-CH$$

$$CH_2-O-\underset{\underset{\displaystyle O^-}{\|}}{\overset{\overset{\displaystyle O}{\|}}{P}}-O-CH_2CH_2-NH$$

The object of the experiments was to incorporate DNP-PE into liposomes in the hope that the haptenic group would be recognised by anti-DNP sera. (*N.B.* Antiserum to small haptenic groups may be generated by immunising animals with heterologous proteins to which the haptenic group is coupled covalently; in this way the antiserum contains antibodies to both the hapten and the carrier protein.) This would open the way to manipulating the chemistry of the antigenic determinants in a controlled manner. As in previous experiments, complement mediated loss of incorporated marker was obtained only under conditions of "active" sensitisation, i.e. when the antigenic phospholipid was present at the time that the lipids were dispersed in water to form the liposomes. It was impossible to sensitise by subsequent additions of the antigen (DNP-PE) to previously formed liposomes. Differences between the antigenic responses of Forssman and DNP liposomes were mostly accounted for by differences in the antisera used: the main active ingredient of commercial anti-Forssman sera being an immunoglobulin of the IgM class while that of anti-DNP serum is exclusively of the IgG class.

Passive Sensitisation. If, instead of using DNP-PE, the monoacyl derivative DNP-lysophosphatidyl ethanolamine was used as the antigen, then it became possible to prepare sensitive liposomes by both active and passive treatments.[31]

$$
\begin{array}{l}
\overset{\overset{\displaystyle O}{\|}}{R-C}-O-CH_2 \\
\quad\ \ HO-CH \qquad \overset{\overset{\displaystyle O}{\|}}{} \\
\qquad\ \ CH_2-O-\underset{\underset{\displaystyle O^-}{|}}{P}-O-CH_2CH_2-NH-
\end{array}
$$

The possibility of passive sensitisation probably arises from the much higher critical micelle concentration of the lysophospholipids (10^{-3}–10^{-5} M) compared with the parent substances (10^{-6}–10^{-9} M). As a result of this there should be present in a suspension of a lyso substance, a sufficient concentration of monomeric material which could exchange and become incorporated in preformed liposomes generated from lipids of other classes.

In passively sensitised liposomes, the antigen will be contained only in the outward facing surface (ES) of the outermost shell of phospholipid, and this gives us the chance of solving the dilemma described earlier, whereby complement damaged liposomes release much more entrapped marker than could possibly be contained in their outermost shell. If passively sensitised liposomes would release significantly less marker than actively sensitised liposomes, then one could argue that the high degree of release obtained from actively sensitised liposomes arises from successive damage to inner shells due to penetration of the antibody and complement proteins through the damaged layers. This is not the case: both active and passively sensitised liposomes release about the same proportion of entrapped marker when treated with antibody and complement under appropriate conditions. It therefore seems most unlikely that marker release from the inner layers depends on secondary antigen-antibody reactions in these regions. More likely is the proposition that the lytic component of the C5b–9 complex can migrate from layer to layer and cause sufficiently extensive damage to release the marker from seven or more shells.

Membrane Damage is not due to Enzyme Action

What is the nature of the membrane damage produced by complement? One of the older ideas is that the increase in ion permeability leading to eventual lysis stems from enzyme catalysed hydrolysis of the membrane lipids. By the use of liposomes, we are in an admirable position to test this hypothesis. One possible procedure is to measure the release of the products of the hydrolysis reaction, but even when radioactive phospholipids were used to enhance the sensitivity of reaction, it was never possible to measure detectable quantities of degradation products, although the liposomes were fully acceptable substrates for phospholipases.[32] In spite of negative findings, it could still be argued that complement damage produces such an extremely limited degree of hydrolysis (possibly a few molecules per fixed immune complex) that even the radiochemical method would be insufficiently sensitive. An alternative course was to use synthetic phospholipid analogues which cannot serve as substrates for phospholipases. Accordingly, Forssman sensitised liposomes were prepared from the phosphinyl ether analogues of phosphatidyl choline[33] in which the fatty acid residues

$$R-O-CH_2$$
$$R-O-CH$$
$$CH_2-P-CH_2.CH_2-N^+-CH_3$$

diether phosphinyl choline

are attached by ether linkage (as opposed to the ester or amide linkage of the physiological compounds) and lack the—C—O—P—ester linkage of the phospholipids. These substances are unable to serve as substrates for any known phospholipases, and moreover, being fully saturated, they are not subject to damage by any of the lipid peroxidation mechanisms. Regardless of this non-reactivity to the enzymes of phospholipid catabolism, liposomes prepared from these substances remained fully susceptible to the actions of antibody and complement and it seems most unlikely that complement lysis of cells arises from an enzyme catalysed degradation of membrane components.

Evidence against the Pore Hypothesis

The other main idea which has predominated among workers trying to find a mechanism for the action of complement is that the lytic complex generates a structured pore which traverses the membrane,[34] and that the dark staining "pits" seen on negative stain electron microscope images of complement damaged cells are the surface expression of these functional holes. If the electron microscopic "pits" are the true fundamental lesions responsible for complement lysis, then they should also be visible on complement damaged liposomes. Reports from different laboratories indicate differing experiences here,[27, 35] but it appears that the typical complement lesions can be reliably visualised by varying the routines involved in preparing the liposomes for fixation and negative staining. A far more fundamental approach is to enquire into the actual properties of these postulated holes. What is their size and structure? Do they relate to the properties of the complement, or to the composition of the membranes in which they are generated? One way of measuring the size of a hole is to determine its filtration properties towards different solutes (see Chapter 3, p. 144) and by working with liposomes one may of course vary the composition of the internal aqueous phase with as much freedom as one can vary the composition of the lipid.

TABLE 1. Loss of glucose and macromolecular markers from complement damaged liposomes of defined composition.[35]

Liposome composition	Marker loss (% of total contained within liposomes)			
	Glucose (180)	HK (102,000)	G6PDH (230,000)	β-GAL (530,000)
PC + 20 μg globoside per μmole PC	50	—	59	48
PC + 4 μg globoside per μmole PC	18·5	—	10	31
PC + zero globoside	0·9	—	1·6	5·5
Sphingomyelin + 20 μg/μmole PC	44	10	—	13
Sheep red cell extract	42	12	—	5

All incubations were for 30 min at 22°. The extent of marker loss has been corrected for the small degree of non-specific release which occurs in the presence of heat inactivated complement serum, and the percentage values refer to the marker release which occurs on incubation with phospholipase C (PC) or saponin (sphingomyelin and sheep red blood cell lipids).

Sphingomyelin and phosphatidyl choline liposomes sensitised with human globoside 1, and liposomes prepared from sheep red cell lipids were loaded with glucose and the following readily detectable macromolecular markers: hexokinase (MW 102,000), glucose-6-phosphate dehydrogenase (MW 230,000) and β-galactosidase (MW 530,000). All three proteins are greatly in excess of the size of haemoglobin (MW 68,000) towards which complement damaged but osmotically protected red blood cells are known to be impermeable (p. 184). The results of some representative experiments[35] are shown in Table 1. When sensitised PC liposomes are treated with antibody and complement, there is an extensive loss of

H

macromolecular material, which roughly parallels the loss of glucose. When the glucose loss is limited by reducing the amount of glycolipid antigen incorporated into the PC liposomes, the loss of the macromolecular markers is reduced as well. Sensitised sphingomyelin liposomes behave differently when challenged with antibody and complement and it can be seen that the extent of leakage of the macromolecular markers is very much less than the loss of glucose. It was impossible to enhance the degree of macromolecule leakage by increasing the amount of antigen incorporated, increasing the amount of antiserum, raising the temperature or prolonging the incubation, all of which manipulations were effective in increasing the extent of glucose loss. The liposomes prepared from crude sheep red cell lipid extracts behaved in a similar way to the sphingomyelin liposomes, which is consistent with the fact that sphingomyelin is the predominant lipid of sheep red cell membranes.

These experiments tell us that the dimensions of the functional lesion depend more on the constitution of the target than on any intrinsic quality of the complement; they also tell us something about the lytic factor itself. It was remarked above that the lytic factor is probably a mobile entity, able to break the barrier properties of seven or more layers of passively sensitised liposomes of both phosphatidyl choline and sphingomyelin. The lytic factor is able to migrate from layer to layer, wreaking damage as it goes, and yet now we see that the damage to one type of phospholipid (sphingomyelin) is selective, allowing glucose, but not hexokinase, to pass. If we may presume that the movement of the lytic factor from layer to layer takes place through the very lesions which it creates, then this observation allows us to place an upper limit to its size, which must be smaller than hexokinase, the smallest macromolecular marker tested. The membrane attack complex of complement (C5b–9) has a molecular weight in excess of 10^6, an order of magnitude greater than hexokinase and even C8 (MW 163,000) would be unlikely to penetrate holes impassable for hexokinase[36]. We conclude that the lytic factor is a peptide cleavage product of the reaction of C5, 6, 7 upon bound C8. Much is known about the interactions of the proteins which collectively comprise complement, yet little is yet understood about its mode of action.

Conclusion
With this last observation we can begin to see that the liposomes, now thoroughly validated as model targets for complement action, enable us to set new questions of protein chemistry which should bring us nearer to its lytic mechanism. Currently, the best description of its mode of action is probably that of a short lived and localised detergent.[23] Detergency could be a quality of an amphipathic peptide, a cleavage product derived from C8 after its reaction with C5, 6, 7 fixed at the membrane surface. In the very close proximity of the membrane this is able to interdigitate with the

lipids and to reduce the forces of attraction between them, even possibly inserting hydrophilic groups into the barrier zone of the bilayer. This idea is consistent with the known facts and should be amenable to further exploration.

Although no clear answers are yet in sight, it is quite true to say that the introduction of the model target cell into this field by Kinsky and his colleagues has made it possible to view a number of initially plausible mechanistic hypotheses in the bright light of day, and generally (in this case) to reject them with certitude. In truth it may be said that this represents a real measure of progress.

CALCIUM AND THE EXPRESSION OF CELLULAR ACTIVITY: IONOPHORES AS DIAGNOSTIC TOOLS IN CELL BIOLOGY

It is in the very nature of cellular differentiation that individual cells have specialised functions. If unicellular organisms may be said to have any functions at all, then these functions are directed solely towards the main-tenance and reproduction of the self. The basic requirements at this level are a competent metabolic process, coupled with the ability to maintain and express genetic information. All the pressures of natural selection operate on the single cell which alone survives or succumbs. In higher organisms differentiated cells, whilst generally remaining metabolically and genetically competent, have individual functions, and it is the or-ganised structure, the body as a whole, which faces the forces of natural selection. A primary feature of higher organisms, which have specialised tissues, is that individual cells must be directed from the outside to ex-press their specialised function. It is the purpose of this section to consider how some messages controlling the activation of cell function may be relayed across cell membranes.

In a very broad sense, cellular activation processes may be divided into two general classes.[37, 38] One of these concerns the control of metabolic processes; as examples of this we have the activation of glycogenolysis in liver, or of lipolysis in adipose tissue. The central requirement for the generation of intracellular cyclic-3′, 5′-adenylic acid (cAMP) as the key step in metabolic activation processes is now widely recognised. The other class concerns very broadly, processes in which there is an expression of cell mobility. This class embraces such diverse cellular functions as move-ment in slime moulds, banding in toad eggs, dispersion of melanin in amphibian skin, muscular contraction, and granule ejection from secretory cells. Nearly all these processes share a common requirement for calcium either in the solution bathing the cells, or within specialised intracellular organelles[40, 41] (e.g. mitochondria and in striated muscle the sarcoplasmic reticulum). A number of these calcium dependent cellular activation processes have been shown to be inhibited by agents which act to increase the intracellular levels of cAMP. These act either to inhibit the rate of cAMP breakdown (the phosphodiesterase inhibitors, such as theophylline and dibutyryl cAMP) or to stimulate the rate of cAMP generation from its precursor ATP by activation of the enzyme adenylyl cyclase. Generally this enzyme is activated through the attachment of cell specific-hormones and synthetic agonists to receptors at the cell surface. Examples of universal activators, which stimulate adenylyl cyclases in a wide variety of tissues include cholera toxin and the fluoride ion, F^-. In many cases, the simple division of activation by calcium and inhibition by cyclic AMP is more complex; for example, the secretion of insulin from the β-cells of the pancreatic islets of Langerhans clearly comes into the category of a calcium dependent function yet it is invariably enhanced by in the presence of theophylline, a phosphodiesterase inhibitor. Turning to the cellular processes of the immune system, we generally see the simpler picture of calcium dependent activation and inhibition by cyclicAMP. The release of enzymes from polymorphonuclear leucocytes, the release of histamine from mast cells and the transformation of lymphocytes are all systems in which cyclic AMP plays an inhibitory role.

DEGRANULATION OF MAST CELLS AND THE SECRETION OF HISTAMINE

The triggering of histamine secretion from mast cells is an excellent example of cellular activation of the second type described above; the interest in this process going far beyond the very real clinical concern for histamine as the mediator of allergic anaphylaxis. Mast cells are widely distributed in the connective tissues of the body, and may be simply obtained in suspension together with a variety of other cell types, by washing out the peritoneum of rats with saline solution. In this preparation they comprise 2–10% of the cell population, and they can be easily purified to homogeneity by density gradient centrifugation.[39] Electron microscopic examination reveals many characteristics which typify cells from secretory tissues (see Fig. 5). The dominating feature is the presence of numerous, dense-staining, close packed granules, each one surrounded by a perigranular membrane. These granules contain the secretory contents of the cell, which includes 5-hydroxytryptamine (serotonin), heparin, histamine, lysosomal enzymes and other pharmacological agents, and secretion occurs by the ejection of granules from the cell, or when they become exposed to the extracellular environment while still being retained within the cell boundary.[40, 42] The ejection and the alteration of granules which occurs on contact with the extracellular fluid is called degranulation and like other secretory processes, this takes place by initial migration of peripheral granules towards the inner surface of the cell,[43] followed by fusion of the perigranular membrane with the plasma membrane.[42, 46] At a later stage, perigranular membranes fuse with other perigranular membranes already fused with the plasma membrane, and it is this process which leads to the genertion of the labyrinthine cavities containing altered granules (from which histamine has been released) which can be seen within the boundary of the cell but which are effectively extracellular (Fig. 5 (b)). This course of events leading to degranulation by membrane fusion is called exocytosis.[47]

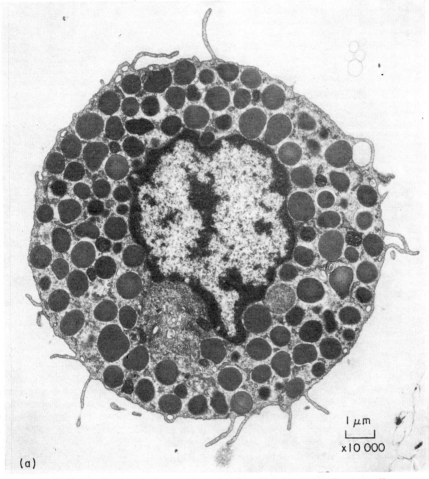

$1 \, \mu m$

x10 000

(a)

Fig. 5. Electron microscope photographs of rat peritoneal mast cells.
(a) An intact cell showing the normal details of cell structure. The nucleus, mitochondria and Golgi region are clearly visible in addition to the profusion of densely stained membrane-bound secretory granules. In fact, this particular cell has been treated with an antibody in order to trigger the secretory process, but only one granule (at 4 o'clock, adjacent to the nucleus) in this section of the cell shows the appearance associated with "release".

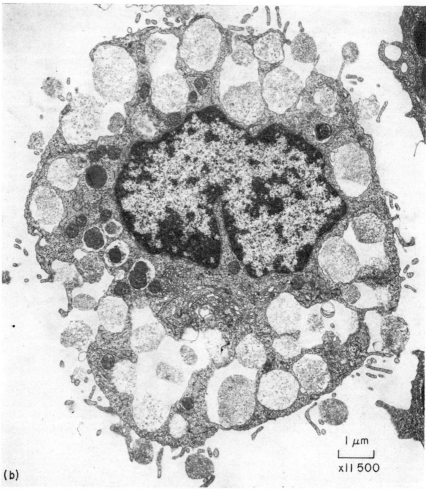

(b)

FIG. 5. Electron microscope photographs of rat peritoneal mast cells.
(b) This extensively released cell was found on the same electron microscope grid as
the cell (a). Only one granule (at 9 o'clock) appears to have survived the release
process. Many granules have been ejected from the cell, while others are retained in
labarynthine cavities exposed to the extracellular environment. The cytoplasm of
the cell is still intact and the structure of the intracellular organelles has been
retained. These cells were fixed with glutaraldehyde, and the sections were stained
with lead citrate. Reproduced by permission of Lawson et al. (1975), J. Exp. Med.
142, 391.

Requirement for Divalent Ligands and Calcium

Histamine can be measured by a number of methods, which include
bioassay,[48] fluorescence[49] and enzyme catalysed radiochemical techniques.[50] There is a wide variety of substances which are able to initiate
the release of histamine from mast cells,[51] but in this account we are con-

cerned exclusively with those agents which elicit secretion by attaching to immunological receptors on the cell surface. These receptors are in reality immunoglobulin molecules of the class IgE[52, 53] which are attached via their Fc portion to Fc receptors in the surface of the cell. The specificity of the surface IgE towards a selected antigenic ligand can be dictated simply by injecting the antigen together with certain adjuvants into the animal about three weeks before the experiment. Alternative ligands which attach to the surface IgE of rat mast cells are antibodies to rat immunoglobulins[54] (e.g. sheep anti-rat immunoglobulin, Santi-RIg) and certain plant lectins such as concanavalin A[55] (conA), which bind to carbohydrate residues associated with the surface IgE molecules. In order to be effective, all these ligands must be divalent (or multivalent);[56, 57, 58] monovalent Fab fragments generated from Santi-RIg preparations, whilst still capable of binding to the surface IgE, are unable to elicit histamine release.[59, 60] The ferritin labelled proteins Santi-RIg-FT and con A-FT retain the capacity to act as triggering ligands, and so one may follow their rate during the stages of exocytosis, by fixation and electron microscopy[46, 60]. It will be recalled that it was by the use of ferritin-labels that the requirement of divalency (or multivalency) was discerned in the phenomena of patch and cap formation by redistribution of surface receptors on lymphocytes and other cells.

Patching (but not capping) is also a consequence of attaching divalent ligands to mast cells, but patching on the scale described earlier is apparently incompatible with degranulation.[60] It does seem possible, though, that a slight degree of surface redistribution, leading to the generation of discrete clusters containing less than 10 IgE molecules (Fig. 6), may be the initial event in the exocytosis process.[60] Exocytosis then commences within a few seconds, and terminates within a minute[61, 62]; it appears to be a self-limiting process in which substantially less than total histamine release and degranulation occurs.[54, 60, 62] Released mast cells are capable of recovery, and the regenerated granules may be secreted following renewed application of a suitable ligand at a later time.[63]

As well as requiring intact respiratory or glycolytic functions,[64] the degranulation of mast cells initiated by immunological ligands has an absolute requirement for an alkaline earth cation in the extracellular medium.[65, 66] Physiologically, calcium is the essential ion, but this may be replaced in *in vitro* experiments by strontium or barium; magnesium is inhibitory.[67] In common with many other calcium dependent secretory processes, histamine release from mast cells treated with divalent immuno-logical ligands, is inhibited by manipulations which raise the intracellular level of cAMP.[68]

Identification of a Second Messenger
The recognition that only calcium among physiological ions when intro-

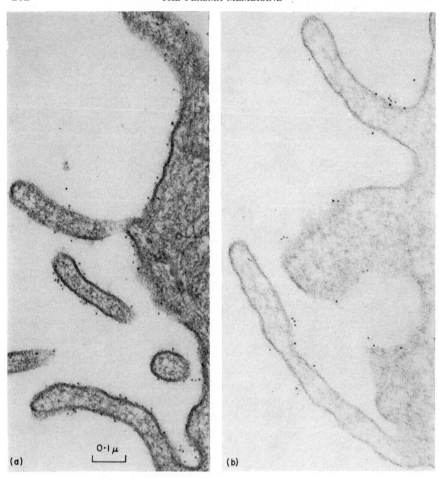

(a)   0·1 μ   (b)

FIG. 6. Rat peritoneal mast cells treated with ferritin labelled ligands.
(a) Treatment with monovalent Fab fragments of anti-rat immunoglobulin serum
for 30 min at 37°. The distribution of the ferritin on the cell surface is random:
observations on a number of cells showed that 90% of the ferritin molecules were
fixed as monomers. A small amount of clustering (8% dimers) could be explained by
the binding of Fab fragments to separate antigenic loci on single IgE molecules on the
cell surface. (b) Treatment with divalent anti-rat immunoglobulin serum for 30 sec
at 37°. Although less of the ferritin-anti-Ig is bound to the cell compared with (a),
there is a marked increase in the degree of clustering. Only 75% of the ferritin
molecules were found to be fixed as monomers, and there was an increase in the
number of clusters containing 2, 3, 4, 5, and 6 ferritin labelled anti-Ig molecules.
Large regions of the membrane are devoid of the marker.

The cells were fixed with glutaraldehyde, and the cell (a) was stained with uranyl
acetate; cell (b) is unstained. Much detailed ultrastructure can be seen in (a); the
"unit membrane" appearance (see Chapter 2, p. 54) can be discerned in regions
where the plasma membrane has been sectioned normal to its plane. Microtubular
and microfilamentous structures are also clearly visible. Reproduced by permission
of Lawson *et al.* (1975), *J. Exp. Med.* **142**, 391.

duced into muscle fibres is capable of inducing shortening[69] forms the basis of a "second messenger" role for this ion. Later with the realisation of an absolute requirement for calcium in other diverse biological activities, it was widely surmised that movement of Ca^{2+} into the cytoplasm is the common trigger for cellular activation.[40] With normal cells of microscopic dimensions this conjecture has always been difficult to prove with any certainty, although attempts to introduce Ca^{+2} into the cytoplasm of individual mast cells by microinjection appeared to support the idea.[70] The aim of the investigation to be described here was to demonstrate that calcium influx is indeed the trigger which initiates degranulation and histamine release, by a process indistinguishable from exocytosis, and to discover some of the ways in which the calcium permeability of the cell is regulated.

In order to prove the role of a second messenger for calcium, we have to show (1) that an inward flux of calcium is the inevitable consequence of fixing divalent ligands to the surface receptors of the cell and (2) that the artificial introduction of calcium into the cell can provoke histamine secretion and degranulation in the absence of attached surface ligands.

The first question was answered affirmatively by the use of the calcium isotope $^{45}Ca^{2+}$. It was shown that there is an enhanced uptake of the radioactive label when mast cells from immunised animals are treated with the specific antigen.[71, 72] As far as is known, the concentration of calcium in the cytoplasm of resting cells is exceedingly low; at 10^{-5}–10^{-8} M it is 2–5 orders of magnitude below that of the extracellular fluid. As a result of this, an increase in the permeability of the cell membrane towards calcium will result in an inward flow, and an increase in the concentration of intracellular calcium.

Inducing Calcium Permeability
It was at this point that the potency of the model approach became apparent. The experimental requirement was for a method of introducing calcium into the cell by a clearly defined route, which does not involve the attachment of external messengers (ligands) to the cell surface, and it was the advent of ionophores for divalent cations which made this possible. It will be recalled (Chapter 3, p. 128) that two carboxylate ion carrier substances, X537A and A23187 have the capacity to transport divalent cations across membranes.[73, 74] When these two substances were applied to mast cells, it became clear that both were capable of eliciting a release of histamine (Fig. 7), but that there are important differences in the quality of the responses.[71]

With X537A it was found that the release of histamine was neither dependent on the presence of Ca^{2+} in the medium, and nor was it prevented by inhibitors of glycolytic or respiratory metabolism. The possibility exists that the release of histamine is a consequence of osmotic lysis due to equilibration of Na^+ and K^+, or due to the direct mobilisation of histamine by this compound, an ionophore of low specificity, capable of carrying monovalent cations and some primary amines as well as divalent cations across lipid barriers.[73]

With the more specific A23187, quite a different situation obtains, and the release of histamine which occurs as a consequence of applying this substance to mast cells more closely resembles the process of exocytosis described above. First, and most important, the release of histamine is

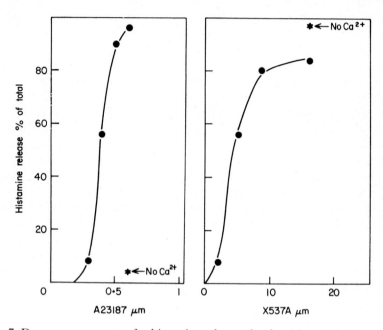

FIG. 7. Dose response curves for histamine release stimulated by application of the ionophores A23187 (specific for divalent cations) and X537A (non-specific) to rat peritoneal mast cells for a 10 min period. The dose response curves are very steep in comparison with dose response relationships commonly obtained for cellular responses resulting from drug-receptor interactions, and it is very unlikely that these curves represent an equilibrium situation of that sort. The steepness of the curves (almost resulting in a switching between nil and total response) may derive in part (for A23187) from the steady state concentration of Ca^{2+} within the cell. When Ca^{2+} is excluded from the medium bathing the cells, A23187 is unable to induce a secretory response, but X537A is insensitive to the presence of Ca^{2+}. With low levels of A23187 it is possible that the cell can maintain the normal low intracellular concentration of Ca^{2+} in spite of the induced influx, by a compensatory efflux due to the pumping action of the Ca-ATPase. Only when this activity becomes overloaded does the Ca^{2+} concentration in the cell rise, and only then does the cell release histamine. This switching effect of A23187 could also be accentuated by the stoichiometry of the Ca^{2+} $(A23187^-)_2$ complex (see Chapter 3, p. 131) which would be expected to result in a square-law dose dependence for the induced Ca^{2+} flux.

strictly dependent on the presence of Ca^{2+} in the medium bathing the cells. Unlike the histamine release mediated by X537A, but like the ligand mediated exocytosis, histamine release due to A23187 is dependent on an intact metabolic process, and can be inhibited by glucose deprivation or blockade of glycolysis or respiration. With this ionophore, it is possible to achieve total release of the mast cell histamine content and in this respect

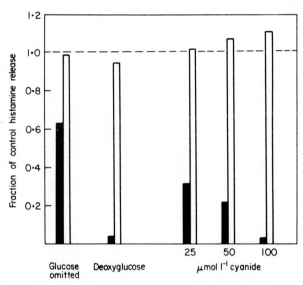

FIG. 8. The effect of glycolytic and respiratory inhibitors on histamine release from rat peritoneal mast cells treated with the ionophores A23187 and X537A. The extent of release, resulting from a 10 min incubation of the cells with the iono-phores is expressed as a fraction of the release obtained in the presence of a normal Tyrode's solution (containing glucose 5·6 mM) and in the absence of inhibitors. Release of histamine induced by the non-specific ionophore X537A (open bars) is insensitive to the effects of glucose deprivation or the presence of deoxyglucose or cyanide. Release induced by ionophore A23187 (solid bars) is sensitive to metabolic inhibition, and in this respect it resembles the secretory process triggered by immunological ligands. Reproduced by permission of Foreman et al. (1973), Nature, 245, 249.

it differs from the ligand mediated process. Furthermore, electron micro-scopic examination of mast cells degranulated with A23187 suggests a degree of structural damage. This is generally revealed as a loss of the ordered array of the membranes of the Golgi apparatus, and the appearance of vacuoles in this region, but it can be more extensive (Fig. 9). Such damage is by no means a necessary consequence of treatment with the ionophore,[75] and many degranulated cells in a representative field closely resemble cells which have been released by normal means. However, in these two respects (total release of histamine, and cellular damage) the secretory process triggered by the ionophore does not mimic exactly the ligand initiated physiological process, in which secretion is followed by recovery.

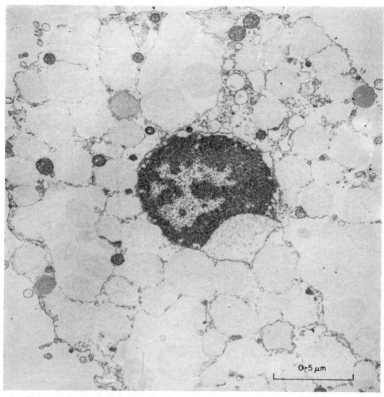

Fig. 9. Rat peritoneal mast cell after treatment with the ion carrier A23187 $(6 \times 10^{-6}$ M) and Ca^{2+} (1·8 mM) for 15 min. There is a considerable degree of structural damage to this cell: the ordered arrangement of the Golgi membranes has been lost and there is extensive vacuolisation in this region. The mitochondria have become peculiarly apparent, and are supported on a structural matrix which is generally devoid of the granular background appearance of cell cytoplasm. The appearance of this cell is not characteristic of a cell which has been released by application of immunological ligands. (Electron microscope photograph by D. Lawson.)

A possible reason for this difference is that the entry of calcium induced by the ionophore is massive, prolonged and uncontrolled. This leads us to the question of how cellular homeostasis is maintained following the normal physiological stimulus to the cell

Desensitisation of the Triggering Mechanism
Not only is there a requirement for calcium in the secretory process, but calcium must be present at the time the triggering ligands are applied to

the cell. If calcium is added to the cell preparation at times after an antigen, then less than the optimal amount of histamine is released (Fig. 10 (a)).[76] The effect of ligands on the secretory process is subject to decay, with the result that there is no response at all if calcium is added 5 min after a triggering antigen. The decay, leading to a null response has been called

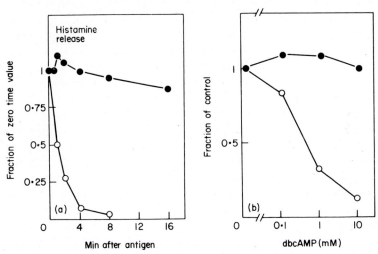

FIG. 10. Differential effects of inhibitory situations on histamine release mediated by an antigen and by the divalent ion carrier, A23187.

(a) The course of desensitisation following application of the specific antigen (egg albumin) to sensitised rat peritoneal mast cells. Addition of Ca^{2+} at times after the antigen results in a progressively reduced secretory response (O—O—O). When Ca^{2+} is added to cells which have previously been treated with the ionophore A23187, the secretory response is maintained over a considerable period of time (●—●—●). Reproduced by permission of Forman and Garland (1974), *J. Physiol.* **239**, 381.

(b) Dose response relationship for inhibition of histamine secretion by dibutyryl cyclic-3′,5′-AMP. The cells were treated with the inhibitor for 30 min, before being triggered with the specific antigen (O—O—O) or with the ionophore A23187 (●—●—●). Reproduced by permission of Foreman *et al.* (1975), *Biochem. Pharmac.* **24**, 588.

desensitisation. The effect of desensitisation may be overcome by application of the ionophore, and we can conclude from this observation that the internal apparatus of the exocytosis mechanism remains intact. It is the ability of the ligand-receptor interaction at the cell surface, to render the membrane permeable to calcium which is lost in the course of desensitisation; the ligand induced calcium permeability pathway appears to open and shut like a gate with a spring. Ligands which act to open the gate will release histamine in a manner which depends on the presence of calcium; substances and manipulations which act to lock the gate will be inhibitory. The calcium carrier molecule A23187 represents an alternative pathway for calcium transport. There is no way in which the cell can restrict the entry of calcium presented by this means, and so normal inhibitors of the calcium gating mechanism are unable to act as inhibitors of ionophore mediated secretion.

Desensitisation of triggered responses is probably widespread. Among other cell processes which are subject to self-limitation by mechanisms which appear to be similar may be included the release of hormones from the neurohypophysis[77] and the adrenal medulla[78] following membrane depolarisation. The kinetics of termination in these examples may best be measured in hours rather than minutes, but the process of calcium gating in mouse T lymphocytes occurs over a very similar time scale to the rat mast cell.[79] Lymphocytes when isolated are dormant; they may be stimulated into DNA synthesis and cell division ("transformation") by lectins and other substances, which show specificity for the two main classes, the T and B cells.[80] These are late events, however, and occur after a delay of 48–72 hours, but in common with the fast responding mast cell, there is a dependence on the presence of calcium[81], and agents which act to raise the level of cAMP are inhibitory.[82, 83] A great deal of interest has centred on early events which may be detected within minutes, or during the first few hours after application of the cell-specific lectin. When the activating lectin, conA, is added to lymphocytes, the T cells take up Ca^{2+}: this has been measured with the use of the radioactive isotope $^{45}Ca^{2+}$, and the induced flux (complete within a minute) is probably the earliest event so far recorded in the sequence leading to transformation.[79] If the radiolabelled $^{45}Ca^{2+}$ is added to the cells 5 min after the addition of the lectin, then this is no longer observed. We conclude that pathways, opened by attachment of divalent ligands to receptors on the cell surface close spontaneously; in this way cells can be protected from the effects of an uncontrolled and unlimited influx of calcium.

Inhibition of Calcium Entry: a Role for Cyclic-AMP

Another situation in which the ligand induced secretory response of mast cells, is inhibited is after treatment with agents which elevate the intracellular level of cAMP. Secretion which results from the mobilisation of calcium with A23187 on the other hand, is not subject to this form of inhibition (see Fig. 10 (*b*)).[84] Situations of this sort, where there is a clear distinction between the effects of inhibitors upon the ligand and carrier induced secretory responses, allow us to point to the mode of action of the inhibitor, and from these findings, we can draw the conclusion that the inhibitory effect of cAMP is expressed at the point of entry of calcium into the cell. The anti-allergic drug disodium cromoglycate ("Intal") also exerts its inhibitory effect on the secretion of histamine at the point of entry of Ca^{2+} into mast cells,[84] but whether this is due to an elevation of intracellular cAMP (cromoglycate has been shown to act as an inhibitor of phosphodiesterase in broken cell preparations[85]) or to a direct effect of the compound acting at the site of calcium entry, is not known. The different sensitivities towards inhibition of the ligand and ionophore mediated release mechanisms may be summarised:

	Ligand mediated secretion	Ionophore mediated secretion	
		A23187	X537A
No Ca^{2+}	inhibited	inhibited	not inhibited
Glycolytic and respiratory inhibitors	inhibited	inhibited	not inhibited
Elevation of cAMP	inhibited	not inhibited	——
Desensitisation	inhibited	not inhibited	——

Once again, direct evidence to support the idea that the inhibitory action of cAMP is exerted by the control of a calcium pathway has come from the study of T lymphocytes in which it was found that prior treatment of the cells with theophylline, dibutyryl cAMP or cholera toxin (all known to elevate cAMP levels) abolished the induced $^{45}Ca^{2+}$ flux due to addition of cell specific lectins.[79] Cyclic AMP locks the calcium gates tight shut.

It should be pointed out that the precise site of action of cAMP (which is likely to act as a co-factor in a protein phosphorylation reaction) has not been delineated: the effect of cAMP on the membrane is most probably indirect, and inhibition of the gating mechanism may result from a sequence of reactions, or cascade. Certainly the generation of cAMP is one possible way by which a cell could act to limit calcium entry, though whether this is the way in which cellular homeostasis is reasserted after a secretory event in normal physiological circumstances (desensitisation) is not known.

Towards the Reality of the Calcium Gate
The calcium gate, as it has been considered so far, is a notional concept based upon the realisation of controlled permeability states of the membrane. Until it becomes possible to isolate a calcium gating substance from biological membranes it will be very difficult to learn much about its nature and mode of operation. In the meantime we have to grasp at such shreds of information as are available to us at the present time. The most promising course so far is to examine the classes of ligands which are able to open the calcium gates and so elicit secretion. As has been pointed out, all the protein-protein interactions which prove effective in triggering the cellular response have the characteristics of polyvalency. Monovalent ligands such as Fab fragments prepared from Santi-RIg (and synthetic

monovalent hapten antigens) can bind to the relevant receptors on the cell surface, but are without effect in the secretory process, and almost certainly this is because they fail to open the calcium gates. Monovalent ligands fail too, in effecting the redistribution of the surface receptors. It would thus appear that the generation of a calcium pathway requires some degree of surface redistribution, and it was suggested earlier (p. 201) that this is probably a precise requirement, involving less than 10 (possibly as few as 2) IgE molecules. There is, of course, nothing to suggest that IgE itself is the gating molecule: the special feature of IgE is to be found in the amino acid sequence of its Fc portion, which ensures its attachment to the Fc receptors of the mast cell membrane. The patching phenomenon shows merely that the immune complexes are subject to redistribution on the cell surface,[60, 86] but we may fairly presume that the Fc receptor molecules are also subject to movement in the plane of the membrane, and that their redistribution, following ligand attachment to IgE, is a necessary step in the generation of the open calcium gate.

Is it possible that we have a model of this process in the "insertion-aggregation" proposal for the mechanism of action of alamethicin and the other excitability inducing antibiotics[87] (Chapter 3, p. 166)? The calcium channel may be a transient and precise arrangement of discrete molecules, probably proteins, drawn together into a coherent channelling formation by the crosslinking of the surface receptors: its disruption into a non-conducting mode could be forced upon it by excessive ligand binding leading to stable but random crosslinking of the gating molecules. Alternative mechanisms for calcium gating will also have to be considered. Amongst these one might suggest that divalent ligand binding could exert an allosteric effect on an enzyme controlled event (with a possible role for cAMP directed regulation?); or, returning to a mechanochemical train of thought, by "uncorking" a persistent channelling structure within the membrane.

Calcium in Other Systems

The introduction of calcium ions into cells by way of the ionophore A23187 has been applied to a large number of systems in which it could reasonably be hoped that intracellular Ca^{2+} is the "second messenger" for an activation process. From many examples, one may mention the use of the ionophore to trigger the activity (secretion and aggregation) of blood platelets[88, 89, 90, 91]; the secretion of insulin[92, 93, 94, 95] and of enzymes[96] from the pancreas; the secretory activity of the salivary glands[97, 98]; of the thyroid[99]; of the rat striatum (dopamine)[100]; of lysosomal enzymes from human neutrophils[101]; of catecholamines from the cat adrenal[102] and of vasopressin from the neurohypophysis.[103] The ionophores for divalent ions have also been used to mobilise calcium in single muscle fibres[104] and to increase cardiac contractility.[73, 105, 106, 107] Among delayed responses which are triggered by the inophore A23187, there is the induction of a parthenogenic reaction in the oocytes of a wide range of species,[108, 109, 110, 111] and the stimulation of transformation (DNA synthesis, blastogenesis and mitosis) in the lymphocytes of pigs and humans.[112, 113] In nearly all of these activities, a role for calcium has been inferred, but it must be pointed out that there do exist some pitfalls which can trip even the most wary investigator. In particular, it is worth noting that three years after the introduction of the ionophore A23187 to cell biology there are no reports of its being successfully applied to the stimulation of DNA synthesis in cultured fibroblasts. Like lymphocytes, fibroblasts are actively investigated in order to learn about the regulation of cell proliferation and growth. In the experimental situation, fibroblasts have the advantage of existing as a homogeneous cell population; progress through the cell cycle may be halted by withdrawing serum factors from the culture medium, and growth can be reinstated by adding these factors back.[114] DNA synthesis, and the subsequent events of transformation are dependent upon the presence of Ca^{2+}, and manipulations which lead to elevation of cAMP are inhibitory.[83, 115, 116, 117] One might have thought that cultured fibroblasts would prove to be the ideal system on which to apply the ionophore A23187 in order to test the hypothesis that calcium entry is the sole and sufficient regulator of cell transformation and growth.

Earlier it was stated that two criteria should be satisfied in order to be sure that it is the movement of Ca^{2+} into the cell which is the second messenger for an activation process. These are (1) that there should be a movement of Ca^{2+} into the cytoplasm, either from storage sites within the cell or (preferably, from the experimental point of view) from the exterior as a consequence of treating the cell with an external messenger and (2) that the artificial introduction of Ca^{2+} into the cell should stimulate cellular activity. To these criteria, we may now add a third: that the triggered movement of Ca^{2+} should be actually relevant to the expression of cellular activity. In the case of the sensitised mast cell, this is demonstrably the case. If calcium is withheld from the cells until after the closure of the gating system, then exocytosis is prevented. With T lymphocytes the situation is different. The permeability of these cells to calcium is temporarily increased following treatment with a triggering ligand (conA)[79] and the transformation (DNA synthesis and mitosis) may be stimulated by direct introduction of Ca^{2+} with the ionophore A23187.[112, 113] However, if Ca^{+2} is withheld from these cells for a few hours after the time of addition of the mitogenic ligand, and then added at a time when the low permeability state has been reasserted, DNA synthesis proceeds as normal.[118] It would appear that it is not the entry of Ca^{2+} into the lymphocytes themselves which finally stimulates them into activity, though it cannot be ruled out that calcium entry into another cell-type present in suspension with the lymphocytes stimulates activity by a cooperative cell-cell interaction.

CONCLUSION

The main theme of this book has been to indicate how a study of simplified models is an entirely relevant and meaningful approach in unravelling the problems of complex biological membranes, their structures and their functional properties. Indeed, it must be apparent that the models (both of membrane structure and of transport systems) are of intense interest in their own right, and that many of the fundmental concepts which we now accept in the field of membrane biology were first revealed by a study of the models without any reference to their biological counterparts. A striking example of this is the concept of lateral fluidity, the description of which in phospholipid membrane models preceded by some years its recognition as a vital feature of the living cell membrane. It was the object of this final chapter to cast the field of interest somewhat wider in order to show that the models (given a fundamental understanding of their properties and limitations) can be applied in an equally valid manner to provide answers to questions in the realm of cell biology. As one goes from the isolated cell membrane, and the simpler cell types with which we were concerned earlier, towards the problems of cellular control, one is so frequently confronted by an impenetrable wall of total confusion. There are observations and apparently conflicting phenomena of uncertain temporal relationship, all wanting to be embraced by a unifying description.

This barrier arises because of the imprecision with which we are able to ask our questions and the bluntness of the pharmacological tools with which we try to probe the functional receptor sites of the cell surface and ascribe defined functions to them. One precise answer to but one well set question can, in such circumstances, have ramifications which go far beyond the immediate question itself. It is rather like the situation of a number of simultaneous equations, the solution of which want for the provision of but one piece of information. This is the role of the membrane model in cell biology—to liberate the science from phenomenological description, by the provision of unambiguous guidelines upon which new interpretations can be built, and new questions asked.

REFERENCES

1.   Ehrlich, P. and Morgenroth, J. (1899). Contributions to the theory of lysin action (in German). *Berlin Klin. Wochenschr.* **361**, 130. *In* "The Collected Papers of Paul Ehrlich (1957), Vol. II", p. 150, F. Himmelweith (ed). Pergamon Press, London.

   * References marked with an asterisk (*) are mainly review articles especially recommended for further reading.

2. *Müller-Eberhard, H. J. (1970). The molecular basis of the biological activities of complement. In "The Harvey Lectures, Vol. 66", p. 75. Academic Press, New York and London.

3. Osler, A. G., Randall, H. G., Hill, B. M. and Ovary, Z. (1959). Studies on the mechanism of hypersensitivity phenomena: the participation of complement in the formation of anaphylatoxin. J. Exp. Med. 110, 311.

4. Henson, P. M. and Cochrane, C. G. (1969). Immunological induction of increased vascular permeability: a rabbit passive cutaneous anaphylactic reaction requiring complement, platelets and neutrophils. J. Exp. Med. 129, 153.

5. Ward, P. A. (1969). The heterogeneity of chemotactic factors for neutrophils generated from the complement system. In "Cellular and Humoral Mechanisms in Anaphylaxis and Allergy", p. 279, H. Z. Movat (ed). Karger, Basel.

6. Jensen, J. A., Snyderman, R. and Mergenhagen, S. E. (1969). Chemotactic activity, a property of guinea pig C′5—anaphylatoxin. In "Cellular and Humoral Mechanisms in Anaphylaxis and Allergy", p. 265, H. Z. Movat (ed). Karger, Basel.

7. Taylor, F. B. and Müller-Eberhard, H. J. (1967). Factors influencing lysis of whole blood clots. Nature, 216, 1023.

8. Lepow, I. H., Dias da Silva, W. and Patrick, R. A. (1969). Biologically active cleavage products of components of complement. In "Cellular and Humoral Mechanisms in Anaphylaxis and Allergy", p. 237, H. Z. Mowat (ed). Karger, Basel.

9. Cochrane, C. G., Unanue, E. R. and Dixon, F. J. (1965). A role of polymorphonuclear leukocytes and complement in nephrotic nephritis. J. Exp. Med. 122, 99.

10. Cochrane, C. G. (1968). Immunologic tissue injury mediated by neutrophilic leukocytes. Adv. Immunol. 9, 97.

11. Alper, C. A., Abramson, N., Johnston, R. B. and Rosen, F. S. (1970). Increased susceptibility to infection associated with abnormalities of complement-mediated functions and of the third component of complement (C3). New Eng. J. Med. 282, 349.

12. Miller, M. E. and Nillson, U. R. (1970). A familial deficiency of the phagocytosis-enhancing activity of serum related to a dysfunction of the fifth component of complement (C5). New Eng. J. Med. 282, 354.

13. Shin, H. S., Smith, M. R. and Wood, W. B. (1969). Heat labile opsonins to pneumococcus: involvement of C3 and C5. J. Exp. Med. 130, 1229.

14. Rother, U. and Rother, K. (1961). On a congenital complement defect in rabbits (in German). Z. Immunitätsforsch, 121, 224.

15. *Müller-Eberhard, H. J. (1975). Complement. Ann. Rev. Biochem. 44, 697.

16. Green, H., Fleischer, R. A., Barrow, P. and Goldberg, B. (1959a). The cytotoxic action of immune gamma globulin and complement on Krebs ascites tumor cells: chemical studies. J. Exp. Med. 109, 511.

17. Green, H., Barrow, P. and Goldberg, B. (1959b). Effect of antibody and complement on permeability control in ascites tumor cells and erythrocytes. J. Exp. Med. 110, 699.

18. Valet, G. and Opferkuch, W. (1975). Mechanism of complement induced lysis: demonstration of a three step mechanism of EAC1-8 cell lysis by C9 and of a non-osmotic swelling of erythrocytes. J. Immunol. 115, 1028.

19. Borsos, T., Dourmashkin, R. R. and Humphrey, J. H. (1964). Lesions in erythrocyte membranes caused by immune haemolysis. Nature, 202, 251.

20. Humphrey, J. H. and Dourmashkin, R. R. (1969). The lesions in cell membranes caused by complement. *Adv. Immunol.* **11**, 75.

21. Mayer, M. M. (1961). Development of the One-Hit theory of immune hemolysis. *In* "Immunochemical Approaches to Problems in Microbiology," p. 268, M. Heidelberger and O. Plescia (eds). Rutgers University Press.

22. Iles, G. H., Seeman, P., Naylor, D. and Cinader, B. (1973). Membrane lesions in immune lysis: surface rings, globule aggregates and transient openings. *J. Cell. Biol.* **56**, 528.

23. *Kinsky, S. C. (1972). Antibody-complement interaction with model membranes. *Biochim. Biophys. Acta,* **265**, 1.

24. Haxby, J. A., Kinsky, C. B. and Kinsky, S. C. (1968). Immune response of a liposomal model membrane. *Proc. Nat. Acad. Sci.* **61**, 300.

25. Six, H. R., Young, W. W., Kei-ichi, U. and Kinsky, S. C. (1974). Effect of antibody-complement on multiple *vs* single compartment liposomes: application of a fluorometric assay for following changes in liposomal permeability. *Biochemistry,* **13**, 4050.

26. Haxby, J. A., Götze, O., Müller-Eberhard, H. J. and Kinsky, S. C. (1969). Release of trapped marker from liposomes by the action of purified complement components. *Proc. Nat. Acad. Sci.* **64**, 290.

27. Hesketh, T. R., Dourmashkin, R. R., Payne, S. N., Humphrey, J. H. and Lachmann, P. J. (1971). Lesions due to complement in lipid membranes. *Nature,* **233**, 620.

28. Kinsky, S. C., Haxby, J. A., Zopf, D. A., Alving, D. R. and Kinsky, C. B. (1969). Complement-dependent damage to liposomes prepared from pure lipids and Forssman hapten. *Biochemistry,* **8**, 4149.

29. Siddiqui, B. and Hakomori, S. (1971). A revised structure for the Forssman glycolipid hapten. *J. Biol. Chem.* **246**, 5766.

30. Inoue, K., Kataoka, T. and Kinsky, S. C. (1971). Comparative responses of liposomes prepared with different ceramide antigens to antibody and complement. *Biochemistry,* **10**, 2574.

31. Uemura, K. and Kinsky, S. C. (1972). Active *vs* passive sensitization of liposomes toward antibody and complement by dinitrophenolated derivatives of phosphatidyl ethanolamine. *Biochemistry,* **11**, 4085.

32. Inoue, K. and Kinsky, S. C. (1970). Fate of phospholipids in liposomal model membranes damaged by antibody and complement. *Biochemistry,* **9**, 4767.

33. Kinsky, S. C., Bonsen, P. P. M., Kinsky, C. B., van Deenen, L. L. M. and Rosenthal, A. F. (1971). Preparation of immunologically responsive liposomes with phosphonyl and phosphinyl analogs of lecithin. *Biochim. Biophys. Acta,* **233**, 815.

34. Mayer, M. M. (1972). Mechanisms of cytolysis by complement. *Proc. Nat. Acad. Sci.* **69**, 2954.

35. Kataoka, T., Williamson, J. R. and Kinsky, S. C. (1973). Release of macromolecular markers (enzymes) from liposomes treated with antibody and complement: an attempt at correlation with electron microscopic observations. *Biochim. Biophys. Acta,* **298**, 158.

36. Kolb, W. P. and Müller-Eberhard, H. J. (1975). The membrane attack mechanism of complement. *J. Exp. Med.* **141**, 724.

37. Allison, A. C. (1973). The role of microfilaments and microtubules in cell movement, endocytosis and exocytosis. *In* "Cell Locomotion 14th Ciba Foundation Symposium" (new series) (p. 109). Ruth Porter and D. W. Fitzsimons (eds). North Holland-Elsevier, Amsterdam.

38. Rasmussen, H., Goodman, D. B. P. and Tenenhouse, A. (1972). The role of cyclic AMP and calcium in cell activation. *Crit. Rev. Biochem.* **1**, 95.
39. Chakravarty, N. and Zeuthen, E. (1965). Respiration of rat peritoneal mast cells. *J. Cell. Biol.* **25**, 113.
40. Douglas, W. W. (1968). Stimulus-secretion coupling: the concept and clues from chromaffin and other cells. *Brit. J. Pharmac.* **34**, 451.
41. Rubin, R. P. (1970). The role of calcium in the release of neurotransmitter substances and hormones. *Pharmacol. Rev.* **22**, 389.
42. Palade, G. (1975). Intracellular aspects of the process of protein synthesis. *Science*, **189**, 347.
43. Uvnäs, B. (1974). Histamine storage and release. *Fed. Proc.* **33**, 2172.
44. Lagunoff, D. (1973). Membrane fusion during mast cell secretion. *J. Cell. Biol.* **57**, 252.
45. Anderson, P., Slorach, S. A. and Uvnäs, B. (1973). Sequential exocytosis of storage granules during antigen-induced histamine release from sensitized rat mast cells *in vitro*: an electron microscopic study. *Acta Physiol. Scand.* **88**, 359.
46. Lawson, D., Raff, M. C., Fewtrell, C., Gomperts, B. D. and Gilula, N. B. (1977). Molecular events during membrane fusion: a study of exocytosis in rat peritoneal mast cells. *J. Cell. Biol.* (in the press).
47. de Duve, C. (1963). The term "Exocytosis" first coined. *In* "The lysosomes", p. 126, A. V. S. de Reuck and M. P. Cameron (eds). Ciba Foundation Symposium. J. and A. Churchill, London.
48. Boura, A. L. Mongar, J. L. and Schild, H. O. (1954). Improved automatic apparatus for pharmacological assays on isolated preparations. *Brit. J. Pharmac. Chemother.* **9**, 24.
49. Shore, P. A., Burkhalter, A. and Cohn, V. H. (1959). A method for the fluorometric assay of histamine in tissues. *J. Phamac. Exp. Therap.* **127**, 182.
50. Snyder, S. H. and Taylor, K. M. (1972). Assay of biogenic amines and their deaminating enzymes in animal tissues. *In* "Research Methods in Neurochemistry, Vol. 1", N. Marks and R. Rodnight (eds). Plenum Press, London.
51. Paton, W. D. M. (1957). Histamine release by compounds of simple chemical structure. *Pharmac. Revs.* **9**, 269.
52. Ishizaka, K. and Ishizaka, T. (1967). Identification of γE antibodies as a carrier of reaginic acitivity. *J. Immunol.* **99**, 1187.
53. Bach, M. K., Bloch, K. J. and Austen, K. F. (1971). IgE and IgG$_A$ antibody-mediated release of histamine from rat peritoneal cells: optimum conditions for *in vitro* preparation of target cells with antibody and challenge with antigen. *J. Exp. Med.* **133**, 752.
54. Humphrey, J. H., Austen, K. F. and Rapp, H. J. (1963). *In vitro* studies of reversed anaphylaxis with rat cells. *Immunology*, **6**, 226.
55. Keller, R. (1973). Concanavalin A, as model "antigen" for the *in vitro* detection of cell-bound reaginic antibody in the rat. *Clin. exp. Immunol.* **13**, 139.
56. Levine, B. B. (1965). Studies in antigenicity: the effect of succinylation of epsilon-aminogroups on antigenicity of benzoylpenicilloyl-L-lysine conjugates in random-bred and in strain 2 guinea pigs. *J. Immunol.* **94**, 111.
57. Magro, A. M. and Alexander, A. (1974). *In vitro* studies of histamine release from rabbit leucocytes by divalent hapten .*J. Immunol.* **112**, 1757.
58. Siraganian, R. P., Hook, W. A. and Levine, B. B. (1975). Specific *in vitro* histamine release from basophils by divalent haptens: evidence for activation

by simple bridging of membrane bound antibody. *Immunochemistry*, **12**, 149.

59. Ishizaka, K. and Ishizaka, T. (1968). Immune mechanisms of reversed type reaginic hypersensitivity. *J. Immunol.* **103**, 588.

60. Lawson, D., Fewtrell, C., Gomperts, B. D. and Raff, M. C. (1975). Anti-immunoglobulin-induced histamine secretion by rat peritoneal mast cells studied by immunoferritin electron microscopy. *J. Exp. Med.* **142**, 391.

61. Bloom, G. D. and Chakravarty, N. (1970). Time course of anaphylactic histamine release and morphological changes in rat peritoneal mast cells. *Acta Physiol. Scand.* **78**, 410.

62. Kaliner, M. and Austen, K. F. (1974). Cyclic AMP, ATP and reversed anaphylactic histamine release from rat mast cells. *J. Immunol.* **112**, 664.

63. Okpako, D. T. (1972). Recovery of anaphylactic sensitivity in isolated guinea pig lungs after densitization. *Int. Arch. Allergy Appl. Immunol.* **43**, 395.

64. Perera, B. A. V. and Mongar, J. L. (1965). Effect of anoxia, glucose and thioglycollate on anaphylactic and compound 48/80-induced histamine release in isolated rat mast cells. *Immunology*, **8**, 519.

65. Mongar, J. L. and Schild, H. O. (1958). The effect of calcium and pH on the anaphylactic reaction. *J. Physiol.* **140**, 272.

66. Diamant, B., Grossman, N., Stahlskov, P. and Thomle, S. (1974). Effect of divalent cations and metabolic energy on the anaphylactic histamine release from rat peritoneal mast cells. *Int. Arch. Allergy Appl. Immunol.*, **47**, 412.

67. Foreman, J. C. and Mongar, J. L. (1972). The role of the alkaline earth ions in anaphylactic histamine secretion. *J. Physiol.* **224**, 753.

68. Parker, C. W., Sullivan, T. J. and Wedner, H. J. (1974). Cyclic AMP and the immune response. *In* "Advances in Cyclic Nucleotide Research, vol. 4", P. Greengard and G. A. Robison (eds). Raven Press, New York.

69. Heilbrunn, L. V. and Wiercinski, F. J. (1947). The action of various cations on muscle protoplasm. *J. Cell. Comp. Physiol.* **29**, 15.

70. Kanno, T., Cochrane, D. E. and Douglas, W. W. (1973). Exocytosis (secretory granule extrusion) induced by injection of calcium into mast cells. *Can. J. Physiol. Pharmacol.* **29**, 13.

71. Foreman, J. C., Mongar, J. L. and Gomperts, B. D. (1973). Calcium ionophores and movement of calcium ions following the physiological stumulus to a secretory process. *Nature*, **245**, 249.

72. Foreman, J. C., Hallett, M. B. and Mongar, J. L. (1975). Calcium uptake in rat peritoneal mast cells. *Brit. J. Pharmac.* **142**, 391P.

73. Pressman, B. C. (1973). Properties of ionophores with broad range cation specificity. *Fed. Proc.* **32**, 1698.

74. Reed, P. W. and Lardy, H. A. (1972). A23187 a divalent cation ionophore. *J. Biol. Chem.* **247**, 6970.

75. Kagayama, M. and Douglas, W. W. (1974). Electron microscope evidence of calcium-induced exocytosis in mast cells treated with 48/80 or the ionophores A23187 and X537A. *J. Cell. Biol.* **62**, 519

76. Foreman, J. C. and Garland, L. G. (1974). Desensitization in the process of histamine secretion induced by antigen and dextran. *J. Physiol.* **239**, 381.

77. Nordmann, J. J. (1975). Hormone release and Ca uptake in the rat neuro-hypophysis. *In* "Calcium Transport in Contraction and Secretion", p.281, E. Carafoli, F. Clementi, W. Drabikowski and A. Margreth (eds). North Holland-Elsevier, Amsterdam.

78. Rink, T. J. and Baker, P. F. (1975). The role of the plasma membrane in the

regulation of intracellular calcium. *In* "Calcium Transport in Contraction and Secretion", p. 227, E. Carafoli, F. Clementi, W. Drabikowski and A. Margreth (eds). North Holland-Elsevier, Amsterdam.

79. Freedman, M. H., Raff, M. C. and Gomperts, B. D. (1973). Induction of increased calcium uptake in mouse T lymphocytes by concanavalin A and its modulation by cyclic nucleotides. *Nature*, **255**, 378.

80. Raff, M. C. (1973). T and B lymphocytes and immune responses. *Nature*, **242**, 19.

81. Alford, R. H. (1970). Metal cation requirements for phytohemagglutinin-induced transformation of human peripheral blood lymphocytes. *J. Immunol.* **104**, 698.

82. Smith, J. W., Steiner, A. L. and Parker, C. W. (1971). Human lymphocyte metabolism: effects of cyclic and noncyclic nucleotides on stimulation by phytohaemagglutinin. *J. Clin. Invest.* **50**, 442.

83. *Abell, C. W. and Monahan, T. M. (1973). The role of adenosine 3',5'-cyclic monophosphate in the regulation of mammalian cell division. *J. Cell. Biol.* **59**, 549.

84. Foreman, J. C., Mongar, J. L., Gomperts, B. D. and Garland, L. G. (1975). A possible role for cyclic AMP in the regulation of histamine secretion and the action of cromoglycate. *Biochem. Pharmac.* **24**, 588.

85. Roy, A. C. and Warren, B. T. (1974). Inhibition of cAMP phosphodiesterase by disodium cromoglycate. *Biochem. Pharmac.* **23**, 917.

86. Taylor, R. B., Duffus, W. P. H., Raff, M. C. and de Petris, S. (1971). Redistribution and pinocytosis of lymphocyte surface immunoglobulin molecules induced by anti-immunoglobulin antibody. *Nature New Biol.* **233**, 225.

87. Baumann, G. and Mueller, P. (1975). A molecular model of electrical excitability. *J. Supramolec. Structure*, **2**, 538.

88. Feinman, R. D. and Detwiler, T. C. (1974). Platelet secretion induced by divalent cation ionophores. *Nature*, **249**, 172.

89. Massini, P. and Luscher, E. F. (1974). Some effects of ionophores for divalent cations on blood platelets: comparison with the effects of thrombin. *Biochim. Biophys. Acta*, **372**, 109.

90. White, J. G., Rao, G. H. R. and Gerrard, J. M. (1974). Effects of the ionophore A23187 on blood platelets: Influence on aggregation and secretion. *Amer. J. Pathol.* **77**, 135.

91. Gerrard, J. M., White, J. G. and Rao, G. H. R. (1974). Effects of the ionophore A23187 on blood platelets: Influence on ultrastructure. *Amer. J. Pathol.* **77**, 151.

92. Wollheim, C. B., Blondel, B., Trueheart, P. A., Renold, A. E. and Sharp G. W. G. (1975). Ca-induced insulin release in monolayer culture of the exocrine pancreas: studies with the ionophore A23187. *J. Biol. Chem.* **250**, 1354.

93. Karl, R. C., Zawalich, W. S., Ferrendelli, J. A. and Matschinsky, F. M. (1975). The role of Ca^{2+} and cyclic adenosine 3':5'-monophosphate in insulin release induced *in vitro* by the divalent cation ionophore A23187. *J. Biol. Chem.* **250**, 4575.

94. Hellman, B. (1975). Modifying actions of calcium ionophores on insulin release. *Biochim. Biophys. Acta*, **399**, 157.

95. Charles, M. A., Lawecki, J., Pictet, R. and Grodsky, G. M. (1975). Insulin secretion: interrelationships of glucose, cyclic adenosine 3':5'-monophosphate and calcium. *J. Biol. Chem.* **250**, 6134.

96. Eimerl, S., Savion, N., Heichal, O. and Selinger, Z. (1974). Induction of

enzyme secretion in rat pancreatic slices using the ionophore A23187 and calcium. *J. Biol. Chem.* **249**, 3991.

97. Prince, W. T., Rasmussen, H. and Berridge, M. J. (1973). The role of calcium in fly salivary gland secretion analysed with the ionophore A23187. *Biochim. Biophys. Acta,* **329**, 98.

98. Selinger, Z., Eimerl, S. and Schramm, M. (1974). A calcium ionophore simulating the action of the α-adrenergic receptor. *Proc. Nat. Acad. Sci.* **71**, 128.

99. Grenier, J. M., van Sande, J., Glick, D. and Dumont, J. E. (1974). Effect of ionophore A23187 on thyroid secretion. *FEBS Letters,* **49**, 96.

100. Holz, R. W. (1975). The release of dopamine from synaptosomes from rat striatum by the ionophores X537A and A23187. *Biochim. Biophys. Acta,* **375**, 138.

101. Smith, R. J. and Ignarro, L. J. (1975). Bioregulation of lysosomal enzyme secretion from human neutrophils: roles of guanosine 3′-5′-monophosphate and calcium in stimulus secretion coupling. *Proc. Nat. Acad. Sci.* **72**, 108.

102. Garcia, A. G., Kirpekar, S. M. and Prat, J. C. (1975). A calcium ionophore stimulating the secretion of catecholamines from the cat adrenal. *J. Physiol.* **244**, 253.

103. Nordmann, J. J. and Currell, G. A. (1975). The mechanism of calcium ionophore-induced secretion from the rat neurohypophysis. *Nature,* **253**, 646.

104. Hainaut, K. and Desmedt, J. E. (1974). Calcium ionophore A23187 potentiates twich and intracellular calcium release in single muscle fibres. *Nature,* **252**, 407.

105. Schaffer, S. W., Safer, B., Scarpa, A. and Williamson, J. R. (1974). Mode of action of the calcium ionophores X537A and A23187 on cardiac contractility. *Biochem. Pharm.* **23**, 1609.

106. de Guzman, N. T. and Pressman, B. C. (1974). The inotropic effects of the calcium ionophore X-537A in the anesthetized dog. *Circulation,* **49**, 1072.

107. Holland, D. R., Steinberg, M. J. and Armstrong, W. McD. (1975). A23187: a calcium ionophore that directly increases cardiac contractility. *Proc. Soc. Exp. Biol. Med.* **148**, 1141.

108. Steinhardt, R. A. and Epel, D. (1974). Activation of sea urchin eggs by a calcium ionophore. *Proc. Nat. Acad. Sci.* **71**, 1915.

109. Chambers, E. L., Pressman, B. C. and Rose, B. (1974). The activation of sea urchin eggs by the divalent ionophores A23187 and X537A. *Biochem. Biophys. Res. Commun.* **60**, 126.

110. Steinhardt, R. A., Epel, D., Carroll, E. J. and Yanagimachi, R. (1974). Is calcium ionophore a universal activator for unfertilized eggs? *Nature,* **252**, 41.

111. Schuetz, A. W. (1975). Induction of nuclear breakdown and meiosis in *Spisula solidissima* oocytes by calcium ionophores. *J. Exp. Zool.* **191**, 433.

112. Maino, V. C., Green, N. M. and Crumpton, M. J. (1974). The role of calcium ions in initiating transformation of lymphocytes. *Nature,* **251**, 324.

113. Luckasen, J. R., White, J. G. and Kersey, J. H. (1974). Mitogenic properties of a calcium ionophore, A23187. *Proc. Nat. Acad. Sci.* **71**, 5088.

114. Holley, R. W. and Kiernan, J. A. (1974). Control of the initiation of DNA synthesis in 3T3 cells: serum factors. *Proc. Nat. Acad. Sci.* **71**, 2908.

115. Balk, S. D. (1971). Calcium as a regulator of the proliferation of normal but not of transformed, chicken fibroblasts in a plasma-containing medium. *Proc. Nat. Acad. Sci.* **68**, 271.

116. Johnson, G. S. and Pastan, I. (1972). Role of 3′-5′-adenosine monophosphate

in regulation of morphology and growth of transformed and normal fibro-blasts. *J. Nat. Cancer Inst.* **48**, 1377.

117. Otten, J., Johnson, G. S. and Pastan, I. (1972). Regulation of cell growth by cyclic adenosine 3′:5′-monophosphate. *J. Biol. Chem.* **247**, 7082.

118. Diamantstein, T. and Ulmer, A. (1975). The control of the immune response *in vitro* by Ca^{++}; the Ca^{++}-dependent period during mitogenic stimulation. *Immunology*, **28**, 121.

INDEX

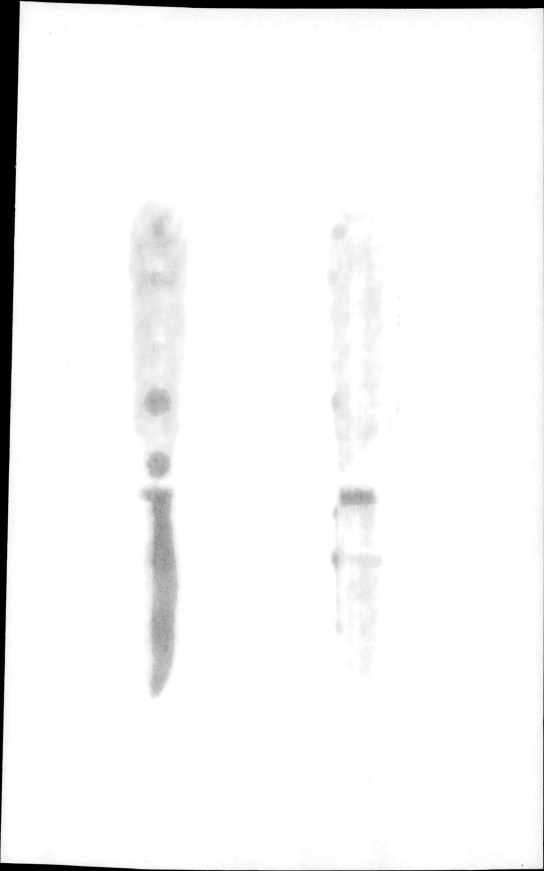